Science and Cultural Theory

A Series Edited by Barbara Herrnstein Smith

and E. Roy Weintraub

AIRCRAFT STORIES

Decentering the Object in Technoscience

JOHN LAW

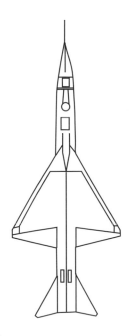

Duke University Press   Durham and London 2002

© 2002 Duke University Press
All rights reserved
Printed in the United States of America on acid-free paper ∞
Typeset in Melior by Tseng Information Systems, Inc.
Library of Congress Cataloging-in-Publication Data appear
on the last printed page of this book.

CONTENTS

Acknowledgments **vii**

1. Introduction **1**

2. Objects **12**

3. Subjects **38**

4. Cultures **65**

5. Heterogeneities **89**

6. Aesthetics **115**

7. Decisions **143**

8. Arborescences **163**

9. Pinboards **188**

Notes **205**

References **225**

Index **241**

The list of persons and organizations who have contributed to this work is too long to include in full. But I would particularly like to mention and thank the following: British Aerospace plc and Rolls Royce plc offered access to material relevant to the TSR2. I am deeply grateful to them for their generous help and assistance without which it would have been impossible to write the present book. This support has come in many forms over a number of years, and has gone far beyond the routine. Accordingly, I thank both organizations for their systematic support and assistance while noting that what I have written is my own responsibility, and does not necessarily reflect the views of either company.

It is also my particular pleasure to thank the Brooklands Museum at Weybridge, Surrey, British Aerospace North West Heritage Group at Warton, Lancashire, and the Rolls Royce Heritage Trust at Filton, Avon. These are organizations, largely staffed by volunteers, that are responsible for collecting and collating the historical records of the two companies. Their work is indispensable to any student of the history of aviation in the United Kingdom and has been crucial in many ways to the present study. I am deeply grateful to them and in particular to the many individuals who, in serving the historical record in this way, have also generously facilitated the study and helped to ease my way at every turn.

I would also like to thank the numerous employees of British Aerospace and Rolls Royce plc and their predecessor companies, and a number of related companies with whom I corresponded. In many cases these people also agreed to be interviewed, and I am particularly grateful to them for generously giving up their time to delve into a project that left distressing memories for many. The same is also true for the politicians, civil servants, and Royal Air Force officers who also unsparingly gave of their time. Since some of them prefer to remain anonymous, I will not here mention any of these kind people by name. In many cases, however, they offered crucial insights into the TSR2 project, the character of military procurement, the nature of defense thinking, and the management of large technological projects.

I am most grateful to the Nuffield Foundation, Keele University, and the École Nationale Supérieure des Mines de Paris for financial, material, and practical support for the research. The Nuffield Foundation generously offered grant aid to support the original research.

Keele University kindly offered sabbatical and other research leave that made it possible to undertake a sustained period of writing. The École Nationale Supérieure des Mines de Paris, and in particular the Centre de Sociologie, offered material support and encouragement throughout. And the Sociology Program of the Research School of Social Sciences at the Australian National University offered fellowship support that provided the blessed respite from the usual commitments that enabled me to complete the manuscript. The research and this book would most certainly not have been completed without the assistance of these four institutions.

I am very grateful to participants in a number of seminars where I was invited to present earlier versions of parts of this text. These seminars took place at Wetenschaps en Technologiedynamica of the Universiteit van Amsterdam, the Netherlands, CRICT at Brunel University, UK, the Department of Social Anthropology at the University of Cambridge, UK, the Department of Sociology at Copenhagen University, Denmark, le Centre de Sociologie de l'Innovation at the École Nationale Supérieure des Mines de Paris, France, the Centre for Social Theory and Technology at Keele University, UK, the Department of Sociology at Lancaster University, UK, Tema T at Linköping University, Sweden, the Department of History and Philosophy of Science at Melbourne University, Australia, and the Area de Innovacaõ Tecnologica Organizacaõ Industrial (ITOI), Programa de Engenharia de Producaõ, COPPE, of the Federal University of Rio de Janeiro, Brasil. The encouragement, support, and critical comments offered at these seminars have been vital to the process of thinking through the arguments that I make here—though I remain conscious of the fact that I have not succeeded in responding to many of the important points raised.

I am deeply grateful to many friends and scholars who have helped, in some cases unknowingly, but more often in the course of extensive and generous discussion, to create the intellectual and political space that has led to this book. I would like in particular to mention Madeleine Akrich; Malcolm Ashmore; Ruth Benschop; Olaf Boettger; Brita Brenna; Michel Callon; Claudia Castañeda; Bob Cooper; Anni Dugdale; Mark Elam; Martin Gibbs; Donna Haraway; Antoine Hennion; Kevin Hetherington; Karin Knorr-Cetina; Bruno Latour; Nick Lee; Celia Lury; Mike Lynch; Ivan da Costa Marques; Maureen Mc-

Neil; Cecile Méadel; Ingunn Moser; Bernike Pasveer; Peter Peters; Andy Pickering; Vololona Rabeharisoa; Paul Rabinow; Vicky Singleton; Leigh Star; John Staudenmaier sj; Marilyn Strathern; Sharon Trawook; David Turnbull; Helen Verran; Steve Woolgar; Brian Wynne; and two anonymous readers for Duke University Press.

To all of these friends and colleagues I am deeply grateful in more ways than can be told. Many of them have been close intellectual friends for many years—and without them the book would never have been written. I would, however, particularly like to mention two people in this list. I am profoundly grateful first to Michel Callon for his many years of acute intellectual friendship, encouragement, and support, and for his conviction that the book is about distribution; and second, to Annemarie Mol for her strong intellectual friendship, the collaborative work that has gone into earlier versions of this book, and her conviction that knowing may be performed as partially connected ontology. I thank them both.

As I write these lines I realize that all these friends share a common indifference to the bounds of disciplinary knowledge and a willingness to take inter- or nondisciplinary intellectual risks. Such risks seem at least as great in the current climate of unremitting academic audit as they ever have in the past, and I am all the more grateful to them for resisting regional restrictions to the character of intellectual inquiry.

Finally I would like to thank Sheila Halsall, Duncan Law, and Angus Law who have lived with and contributed to this book in one form or another for more than ten years.

A version of chapter 6 has been published in *Configurations,* vol. 8, no. 1 (winter 2000). I am grateful to the publisher for allowing it to be included in this volume.

A plateau is always in the middle, not at the beginning or the end. A rhizome
is made of plateaus. —Gilles Deleuze and Félix Guattari, *A Thousand Plateaus:
Capitalism and Schizophrenia*

No doubt Deleuze and Guattari have got the right idea. Matters grow
from the middle, and from many places. But one also has to start
somewhere.

With the aircraft? *This is a book about specific episodes in a British
attempt to build a military aircraft,* a tactical strike and reconnaissance warplane, called the TSR2. The project to build this aircraft
started in the 1950s and ended in 1965 when it was canceled by a
newly elected Labour government. In one way or another, all the
stories in this book have to do with the TSR2.

But the aircraft is not the only possible place to start. For though
all the stories in this book are indeed about the TSR2, the book is
really about something much more general. *It is about modernism
and its child, postmodernism—and about how we might think past
the limits that these set to our ways of thinking.* For the book is about a
world, the contemporary Euro-American world, in which many have
lost their faith in big theories or "grand narratives," as Jean-François
Lyotard calls them (1984b). And, at least to some extent, it is about a
world in which many have also lost confidence in the grand projects
and plans that tend to go with those grand narratives. Nuclear power,
medical practices, food safety, the environment, everywhere, or so
the story runs, experts are doubted, and people are skeptical of the
claims made by authorities. Including academic authorities.

Of course there are various ways of responding to this. One can
wave aside the skepticism of postmodernism and insist that experts—
including academic experts—still know best: that it is, indeed, possible to tell grand narratives. One can, in short, remain a modernist. Alternatively, one can insist that expert knowledges are limited
in scope, but then go on to say that it is still possible to tell consistent stories so long as one understands that these have only a limited validity and that they will in due course require revision. No
doubt this is the dominant response in many of the social sciences,
for instance underpinning the theory of reflexive modernity.[1] It is a
response that says warrantable knowledge is still possible so long as

it is suitably set about with health warnings and it is not used after its sell-by date.

But there is another possibility that I want to explore in this book. This is to take the skepticism of the so-called postmodern condition seriously, which means accepting that "modernism" is flawed even in its more supple versions. It is to accept that modernism never achieved the smoothnesses it sought, that its foundations were illusory, and that when it intervened to try to put things right and make a better world it often—as Zygmunt Bauman has so eloquently shown—wreaked havoc.[2] But then it recognizes, and this is crucial, that the pluralist diaspora apparently favored by postmodernism raises problems that are just as difficult. Not only is it clear that we don't live in a pluralist world in which everyone happily does their own thing, but it is also apparent that the broken fragments celebrated in postmodernism are just as much a product of modernism as its own streamlined coherences ever were. Postmodernism is, so to speak, the mirror image of modernism—and postmodernism's response has simply been to break the smoothness and shatter that mirror. The argument, then, is that modernism and postmodernism exist together. They are each other's creatures. And as they confront one another they tend to press us to make a choice between the homogeneities of centered storytelling on the one hand, and pluralism of fragmentation on the other. This, then, is a second version of what the book is about. It is an attempt to evade that choice.

But to make the argument I need to be more specific. So a third and more concise way of talking about the stories assembled in this book is to say that they are about *fractional coherence.* Fractional coherence, I will say, is *about drawing things together without centering them.*

Knowing subjects, or so we've learned since the 1960s, are not coherent wholes. Instead they are multiple, assemblages. This has been said about subjects of action, of emotion, and of desire in many ways, and is often, to be sure, a poststructuralist claim. But I argue in this book that *the same holds for objects too.* An aircraft, yes, is an object. But it also reveals multiplicity—for instance in wing shape, speed, military roles, and political attributes. I am saying, then, that an object such as an aircraft—an "individual" and "specific" aircraft—comes in different versions. It has no single center. It is multiple. And yet these various versions also interfere with one another and shuffle themselves together to make a single aircraft. They make what I will call *singularities,* or *singular objects* out of their multiplicity.[3] In short, they make objects that cohere.

But how do they do this? This is the major question that I tackle in this book. A question that, while speaking to the general issue raised by the so-called postmodern predicament, at the same time much more concisely refuses the pluralism implied by Lyotard's multiple language games.

How, then, to think about this? I deploy a range of metaphors for thinking about the overlaps that produce singularity out of multiplicity. Many of these have grown up in the discipline of STS—of science, technology, and society. Interference, oscillation, Donna Haraway's notions of "the established disorder" or the cyborg—these terms catch something important about the relations between singularity and multiplicity. But let me mention a further possibility here, that of fractionality. In mathematics fractals are lines that occupy more than one dimension but less than two.[4] If we take this as a metaphor without worrying too much about the mathematics, then we may imagine that fractal coherences are coherences that cannot be caught within or reduced to a single dimension. But neither do they exist as coherences in two or three separate and independent dimensions. In this way of thinking, a fractionally coherent subject or object is one that balances between plurality and singularity. It is *more than one, but less than many.*

I want to suggest that Euro-American culture doesn't really have the language that it needs to imagine possibilities of this kind. Its conditions of possibility more or less preclude the fractional. Indeed this is one of the reasons why the postmodern reaction—though it diagnoses some of the problems of modernism well enough—still finds itself trapped within a version of the modern predicament. For if things don't cohere together to form a consistent whole, then it is usually assumed that they don't cohere at all. So in common sense (as well as much academic and political discourse) the options tend to take the form of the binarism mentioned earlier: between, on the one hand, something that is a singularity because it holds together coherently; and, on the other, something that is broken and scattered, as in some kind of pluralism in which anything goes.[5] Or between order and its antithesis, chaos. Thus our languages tend to force us to choose between centers or dislocated fragments. Between the poles of "drawing things together" and "the decentering of the subject."[6] Or between single containers, such as "society," and plural elements, such as "individuals," that are contained within society. Fractionality, then, is one of the possible metaphors for trying to avoid such dualisms. For trying to wrestle with the idea that objects, subjects, and societies are both singular and multiple, both one and many. Both/and.

This, then, is the hope: that after the dualist contraries of centering and decentering, after the alternates of singularity and multiplicity, we might find ways of imagining fractionality. This, to be sure, is the hope of a number of scholars and is certainly one of the lessons that we learn from parts of poststructuralism.[7] But the program, it seems to me, has not yet found good ways of performing itself—and least of all of doing so empirically. This leads to the fourth significance for the stories that I tell in the book. A fourth way of beginning.

This starts with a question: *How should we write?* How might we write about multiplicity in a way that also produces the effects of singularity? Or about singularity in a way that does not efface the performances of multiplicity? In this book I do not respond to this question by offering a single recipe or a formula. Instead I choose to proceed less directly and more allegorically. Or, more precisely, I try to make something, to *create* it rather than simply telling about it. For this book explores complexity, heterogeneity, and interference not simply by talking about them, but also, and maybe more importantly, by trying to perform them.

I believe that if we have not managed to attend very well to the fractional coherences of multiple objects and subjects, this is not simply because we have not properly faced the facts. It also has to do with how we investigate our subjects and objects and, in particular, with the ways in which we tell about them. It has, in short, to do with the character of social-science writing. Notwithstanding work in several social-science traditions, we are, to use a phrase, insufficiently self-reflexive about the way in which we write.[8] And about what is implied when we write in one way rather than another. So my hypothesis is that *we have not yet recognized and allowed the difficult subjectivities that are needed for fractional knowing.* In this book I also help to bring such less direct ways of knowing into being. The book, then, is an intervention, a performance of fractional ways of knowing.

Perhaps it would have been possible to make a grand narrative about decentered and yet coherent objects. I take it that this is one of the features of Andrew Pickering's work on the "mangle of practice," a metaphor that otherwise does work which has much in common with what is attempted in this book: an inquiry into ontology, into what is made, rather than what is represented.[9] And the thought of working in terms of a single metaphor is attractive because it offers a key to complexity. And such keys, once in place, are easily expressed and applied. Telling directly about what they tell, they are rendered easily transportable. To say it quickly, such is the dream of modernism in its search for foundational (or now postfoundational) grounds, and it is certainly the project of much contemporary social theory, to which the possibilities of allegory are foreign.[10] But here I explore a less direct alternative by growing different stories alongside one another. Smaller narratives—a lot of smaller keys. Working in this way has a cost: we do indeed lose the possibility of an overall vision. But at the same time we also create something that was not there before: we create and make visible interferences between the stories. We bring new and unpredictable effects into being, effects which cannot be predicted or foretold from a single location. New forms of subjectivity.

To do this is to alter the character of knowing and writing. It is to render them multiple, decentered, or partially centered, in this place that refuses both modernism and postmodernism. If single accounts offering single keys make arborescences—treelike structures with beginnings, middles, and ends where everything important is held together in a centrally coordinated way—then multiple storytelling makes rhizomatic networks that spread in every direction. They make elaborations and interactions that hold together, fractionally, like a tissue of fibers.[11] This results in texts that are uncentered, texts that are not singular. And yet, if the bet is right, it produces texts that have intersections, that hold together. That cohere.

So what does all this mean in practice? The answer is that the essays in this book tell specific stories about specific events. In doing so, they play upon recurrent themes to do with partiality, fractionality, interference, and collusion, while doing so in a manner that resists the simplicities of an overall beginning, middle, and end. The book as a whole, then, is not treelike in structure. It is not an arborescence. Instead it takes the form of a rhizomatic network. It makes overlaps and juxtapositions, and it makes interference effects as a result of making these overlaps. So that is the fourth way of introducing the book. *It is about writing fractionally.*

But this suggests a fifth way of talking about the stories of the book, which has to do with *how texts relate to the world.* Perhaps, to be simple, we might speak of two possibilities. First, we may imagine that they tell about and thus represent a version of reality. If we think of writing in this way, then we distinguish between texts on the one hand, and what they represent on the other. The latter become something separate, out there, prior, removed. This means that we may stand outside and describe the world, and that when we do so we do not get our hands dirty. We are not in the world.

The alternative is to imagine, reflexively, that telling stories about the world also helps to perform that world. This means that in a (writing) performance reality is staged. And such a staging ensures that, everything else being equal, what is being performed is thereby rendered more obdurate, more solid, more real than it might otherwise have been. It becomes an element of the present that may be carried into the future.

So what do we perform when we write? There are various by now familiar possibilities. We may perform the world as a treelike structure: such is the desire of modernism as it seeks to perform its centered consistency into being. We may make fragments, which is, to be sure, the postmodern response. Or we may enact it rhizomatically, which is the allegorical or poststructural alternative that I am recommending.

In this alternative approach, no matter how stories are told about this aircraft, the TSR2, they do not simply describe something that happened once upon a time. They are rather, or also, a way of helping to perform the aircraft. The stories *participate* in the aircraft. They add to the crowd of forms in which it was already among us, interfering with and diffracting earlier versions and thereby altering these forms. Perhaps slightly and locally. Perhaps unpredictably. But nevertheless altering them, and making a difference.

So the performativity of writing is a fifth way of introducing the book, of describing the significance of its stories. But this in turn suggests a sixth possibility: that the book is about *what it is to criticize,* analytically and politically. Its fractional object is, as I have noted, a military aircraft. Why this should have been so is something that I explore in chapter 3. As is obvious, there is much to worry about in military aviation. Had the TSR2 ever been used in its nuclear role, the world would have stumbled into Armageddon. And, leaving aside the horrors of destruction, in the stories that follow we'll come across ways in which the TSR2, even if it never killed, indeed performed social distributions—for instance those of gender or ethnicity.[12] So yes, there is much to worry about here. But there is a problem if we start to criticize from what is supposed to be the outside because doing so ignores the performative character of storytelling that I have just been describing. In particular it ignores the fact that we are all mixed up in what we are describing. That, indeed, in one way or another we are helping to bring it into being. The fact that we are colluding with what we are describing, colluding to enact it into being. The conclusion is that in a fractional and reflexive world the luxury of standing outside, criticizing, and correcting is no longer available. Partly inside, partly outside, we are at least partially connected with our objects of study. And if we seek to criticize then it also becomes important to reflect on the character of that involvement. We need to ask whether, and if so how, we share in what we do not like with those whom we do not like. And whether, and if so how, they share some of our own most valued ways of being.

This should not be misunderstood as a plea for political quietism. Indeed, quite to the contrary. Thus if our writings perform reality, then they also alter it. Every time we act or tell, we also, at least putatively, make a difference. We *always* act politically. The only question is how do we do it?

This book interferes in a variety of ways, but in particular, or so I hope, it interferes with what we might think of as "project-ness." This is the idea (which is also a performance) that many technologies and other social arrangements are properly narrated and organized as "projects," "programs," "operations" or other closely related terms such as "organization," "system," "network," or even the "reflexive person." These are objects that are somewhat linear, chronologically chained, and more or less centrally and teleologically ordered, and that are also shaped in one way or another by their circumstances. Think of the TSR2 project. Or the Manhattan Project. Or the mission statements of organizations. Or indeed "the modern project." Think of large technical systems, or actor-networks. This kind of telling and performing is a standard narrative trope in late modernity. And it is, of course, performative of that modernity, tending as it is told and enacted to order social relations in an image of projectness. It is one of the aims of this book to interfere with this trope, to erode the assumptions performed in projectness, or at least to explore what is involved in their enactment. Thus, the sixth argument of the book in effect suggests to social scientists that, insofar as they frame what they tell in the form of stories about projects, they too are colluding in reproducing the conditions of projectness as an appropriate narrative form. No doubt this is not all bad. There are moments for this collusion. But if the arguments I am making carry any weight, then that performance tends to efface not only other possibilities but also the fractional conditions of the performance of singularity. And, to be sure, set limits to the conditions of possibility.

So there are at least six possible introductions, six ways of telling what the book is about: it is about an aircraft; it is about refusing the space provided by the division between modernism and postmodernism; it is about fractional coherence; it is about the reflexive forms of academic subjectivities needed to apprehend the fractional; it is about the performativity of writing; and it is about the collusions that necessarily follow from that performativity. Such are the themes that recur and interfere with one another throughout the book.

Each of the eight chapters that form the body of this book tells its own story and mobilizes its own resources, drawing variously on cultural studies, technoscience studies, feminist theory, philosophy, sociology, cultural anthropology, art theory and history, and semiotics.

Chapter 2 concerns the problem of multiplicity. It uses a version of semiotics to analyze how an aircraft sales brochure generates first a range of object positions and then coordinates them into a single aircraft. This analysis implies that coherent and single objects are *effects* or products. It also implies a shift from epistemology to ontology. This is because inconsistency between different performances reflects failing coordination between different object positions rather than differences between external perspectives on the same object. These, then, are two of the implications if we start to imagine that narratives are not about self-evidently singular objects but rather have to do with the enactment of fractional relations.

Chapter 3 deals with subjectivity, interpellation, and collusion. It describes how I was multiply interpellated by the TSR2, which implies that there is no such thing as a centered subject: like objects, subjects of knowledge are multiple or fractionally coherent. It also suggests that the interferences between these different subject positions are a valuable source of data. This means that if it is properly used, "the personal" is not confessional but analytical in character. It also, however, means that when subjects are interpellated by objects, they are liable to find themselves colluding in the performance of certain narrative forms. Such was certainly so in the case of the TSR2.

Chapter 4 is about bias in favor of narrative continuity, and the ways in which discontinuities are effaced or deferred. In this chapter I identify three versions of narrative continuity: the chronology of genealogy and descent; the synchronicity of systematic connection; and depth hermeneutics, for instance in the form of background factors such as social interests that then shape more superficial phenomena. Despite their differences (and these, of course, have been rehearsed *in extenso* in social theory), each version performs a bias in favor of continuity and connection, while discontinuities are deferred into slippages between the different narrative forms and so tend to be effaced. This analysis implies that the difference between insider and outsider cannot be sustained: social scientists and participants alike tell their stories in terms of these narrative possibilities. They collaborate to perform projectness and its conditions of possibility, which include a homogeneous space-time box with its own set of coordinates in the form of chronology and scale.

Chapter 5 concerns oscillation between singular presence and multiple absence. It considers an aerodynamic formalism that seeks to draw things together in an explicit and homogeneous manner. This formalism operates by simplifying and excluding almost everything —including other realities that are represented in algebraic form but cannot possibly appear on a sheet of paper. The formalism is thus oscillatory: it necessarily makes absent that which it also seeks to make present. The paradox is that presence and coherence rest on their converse, that which cannot be made present and coherent. This means that absence and presence cannot be dissociated. Again, then, the underlying theme of the chapter is that objects are not singular, indeed not self-identical. That in their heterogeneity they are instead fractional and can only be apprehended fractionally.

Chapter 6 is also about oscillation, this time oscillation first between text and pictures and second within the pictures themselves. The text of the brochure discussed in chapter 2 creates an aircraft that is practical, technically efficacious, and militarily invulnerable. The illustrations extend the performance of military invulnerability but also stress the nonpractical fact that to fly this aircraft is thrilling for a certain kind of heroic male subject. There are other genderings at work as well within the pictures. Though the aircraft itself is sometimes performed as a potent male, there are moments when it is made female in a version of the patriarchal fear of the power of woman performed in the oscillation between Madonna and whore. Thus the aesthetics of the illustrations (themselves noncoherent) interfere with the text in ways that are discursively illegitimate in order to perform a singular and obdurate aircraft that is strong and deadly.

Chapter 7 is about decision making. It explores the assumptions about decision making in descriptions about the decision to cancel the TSR2. These include distinguishing between reality and fantasy; effacing the microphysics of power; performing certain places and times as discretionary; distinguishing between that which is important and that which is a mere "detail"; and (in a further example of the oscillation between singularity and multiplicity) the erasure of differences between different decisions in a framing assumption that the decision taken was indeed one rather than many. This assumption of singularity thus makes it possible for different individual decisions to be made—but, I argue, it is *necessary* for different decisions to be made if a single decision is to be achieved. These, then, are narrative collusions to do with decision making not unlike those entailed in studying "projects." Again there is oscillation.

Chapter 8 returns to narrative performativity and collusion. It offers several accounts of the TSR2 project that reveal substantial overlaps. In particular, it suggests that the accounts are arborescent in form.

Thus the stories all join in the performance of a single TSR2 and its projectness—and the work of building the kind of homogeneous space time box described in chapter 4. This analysis suggests, once again, that that the distinction between insider and outsider doesn't really work; that all accounts are performative (there is a discussion of Austin's performatives and constatives); and that all collude in the reproduction of the conditions of possibility, which include a singular world and a singular object in which the oscillation with multiplicity is effaced. The hands of the storyteller are never clean.

Chapter 9 considers what comes after centering—for, given the gravitational pull of centered storytelling within the narrative traditions of modernism, escaping from singularity is difficult. Indeed, to talk of "escape" is not the right metaphor because it implies a postmodern fragmentation with the binarisms from which we need to escape. In this chapter I first consider the metaphor of the pinboard, the relationship between narratives or other performative depictions juxtaposed on a notice board. I suggest that this metaphor may help us to handle the performative character of our own ways of knowing in a manner that does not conceal their multiplicity. I then return to the question of the political. The question is, does an insistence on fractionality rather than the singularity of social structure imply political quietism? I argue that this is far from the case. Even leaving aside the often-collusive performativity of singular narrative, I suggest that the great social distributions familiar to sociologists and political commentators are all the more obdurate precisely because they are not singular but rather fractional in character. There is no "weak link" in an otherwise coherent structure. Rather there are partial and supple connections between distributions that help to secure dominance and reproduce the established disorder.

All of which—and this is the concluding thought—also demand fractional ways of knowing; skepticism about viewpoints that try to perform themselves as simply centered; and an ability to live and know in tension. This is one version of what a rigorous and politically interventionary social science that seeks to avoid both modernism and postmodernism might look like.

It was a sales brochure. About sixty pages long, it was published in 1962 by the British Aircraft Corporation. And it was trying to sell an aircraft, the TSR2, to its readers. But what *was* the TSR2? And who were the readers of the brochure?

There are historical responses to both these questions. TSR2 was a tactical strike and reconnaissance warplane being designed and built by the major UK aerospace manufacturer, the British Aircraft Corporation. And the brochure was intended for an elite readership: senior air force officers in the UK or in certain "friendly" countries, most notably Australia; senior civil servants, again in these selected countries; and no doubt a number of well-placed politicians. For the brochure was part of an effort to sell the aircraft, both in Britain but more particularly (since the Royal Air Force was already committed to its purchase) to possible overseas buyers.

Those, then, are brief versions of the historical answers. I offer them at the outset because I do not want to be accused of playing games, of withholding context, or of denying the obvious. But the direction in which I wish to move is different. For reasons that will become apparent I do not want to frame what I write in terms of the conventions of narrative history. Though this strategy, of course, brings its costs, I want instead to create a *naive reader*—a naive reader who knows nothing about the TSR2 or the potential readers of the brochure. And I want to use this fiction in order to learn something about how the brochure works. So the thought experiment is this: that we read excerpts from the brochure without making too many assumptions about its character, about what it is telling us, or about its likely readers. Something that is not possible if we arrive with the competences and the concerns of the historian.

So what happens if we do this?

Naive Readings

Exhibit 2.1 is from page twenty-five of the brochure. As is obvious, this is a drawing, the drawing of an aircraft. Then the question arises immediately: how naive do we want to make the reader? If we insist on a radical version of naïveté then we need to say that there is nothing about the picture that links it with the TSR2. For yes, it is a picture of an aircraft. But there is no caption to say that this aircraft is the TSR2.

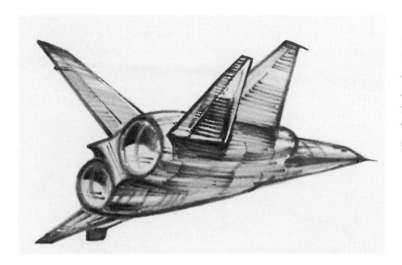

EXHIBIT 2.2 "The T.S.R.2 weapons system is capable of a wide range of reconnaissance and nuclear and high explosive strike roles in all weathers and with a minimum of ground support facilities." (British Aircraft Corporation 1962, 4)

EXHIBIT 2.3 "In T.S.R.2, high grade reconnaissance is allied to very accurate navigation and this suggests the application of the aircraft to survey duties. In many areas the navigation accuracy of better than 0.3% of distance travelled is a significant improvement on the geodetic accuracy of existing maps. This degree of precision enables new maps to be made or old ones to be corrected with a minimum of accurately surveyed reference points." (British Aircraft Corporation 1962, 17)

Exhibit 2.2 appears much earlier in the brochure—indeed on the first full page of text. Here we don't learn anything about an aircraft. Instead, we learn that the TSR2 is a weapons system. We also learn that this weapons system fulfills a range of roles, and that it does so in ways that are independent of the effects of weather and elaborate ground-support facilities. But is it an aircraft? Again, to be sure, it depends just how naive we want to be. But if we were to dig in our heels then we would have to say that we've learned that the "TSR2" is a weapons system, but not that it is an aircraft.

Exhibit 2.3 tells us something about TSR2 and navigation. Here the naive reader does indeed learn that TSR2 is an aircraft, but that reader also learns something about the character of this aircraft: that it has

EXHIBIT 2.4 "In T.S.R.2 the internal and external communications facilities are completely integrated. Two control units provide for intercommunication between the crew and for control of the radio equipment installed." (British Aircraft Corporation 1962, 29)

EXHIBIT 2.5
Fuel System (British
Aircraft Corporation
1962, 41;
© Brooklands
Museum)

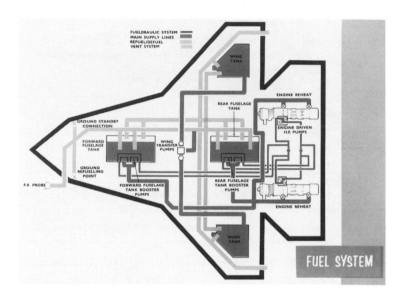

to do with remote sensing and surveying. TSR2, or so it is being suggested here, is an aircraft capable of accurate navigation—but also, and perhaps more remarkably, one that is capable of making maps.

How many more versions of naïveté do we need? Exhibit 2.4 turns the TSR2 into a communications system. Exhibit 2.5 (though, like the drawing in exhibit 2.1, it does not mention the TSR2 by name), turns it into a fuel system, complete with pipes, tanks, pumps, and engines. And exhibit 2.6 (again we need to enter the caveat about the absence of a name) turns it into a global traveler, moving to and fro between Britain, Australia, and a host of other points around the globe.

Let's stop the experiment now. We could pile up more exhibits, but we have learned what we need to learn for the moment: a naive reader who does not start out with an idea of what it is, this TSR2, who does not make connections, will learn that it is many and quite different things.[1] Let me stress the point: "the TSR2" is not a single object; neither, whatever the exhibits might suggest, is it many different parts of a single object. Instead it is many quite different things. It is not one, but many.

This is the *problem of difference:* we have different objects. Or it is the problem of multiplicity: we have multiple objects. In other words, a reader who insists on being naive is likely to find that he or she is dealing not with a single object but rather with an endless series of *different* objects, objects that carry the same name—for instance "TSR2"—but which are quite unlike one another in character.

Of course, we know that it is not really like that. We know—or at least we assume—that the object, the TSR2, is indeed an object. But *why* is this? Why do we make this jump? And how does it come about? The ability to pose such questions is the reason for avoiding a historian's sensibility and the justification for being naive. An initial assumption of naïveté enables us to ask why the reader for whom the brochure was intended would assume that it was, indeed, describing a single object, a single aircraft, rather than a whole flock of different machines. In other words, an initial assumption of naïveté is a methodological position.[2]

But why be naive? To answer this question I need to talk of *strategies of coordination.* In particular, I will identify a series of mechanisms that work to connect and coordinate disparate elements. The

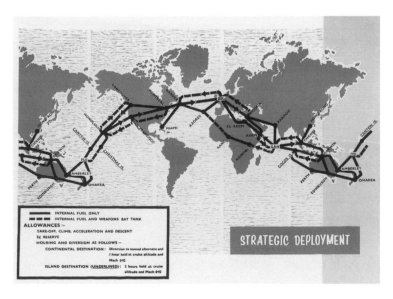

EXHIBIT 2.6 Strategic Deployment (British Aircraft Corporation 1962, 23; © Brooklands Museum)

The Problem of Difference

Annemarie Mol has written a book about this, about the problem of difference in medicine.[3] Think, she says, of lower-limb arteriosclerosis. Or better, think of the *practices* within which lower-limb arteriosclerosis is located. Perhaps we may number three of these.

First, there is a phenomenon the doctors call "claudication." Claudication is suffered by patients. It is pain in the legs occurring when the patient walks further than a certain distance. This is diagnosed in general practitioners' surgeries when the patient is interviewed.

Second, there is the phenomenon of an inadequate flow of blood to the legs and the feet. This usually arises initially in outpatient clinics. The investigating physician measures the pressure of the blood flow at the ankle and compares it with the pressure at some other convenient point such as the top of the arm. If the difference is large then there is said to be pressure loss at the ankle. This loss of pressure is taken to be a sign of increased resistance to the flow of the blood.

Third, there is the phenomenon of the thickening of the intima of the blood vessels in the leg. There are various practices for exploring this, but the most important is located in the pathology laboratory, after the amputation of a diseased leg. The pathologist cuts cross sections through the blood vessels of the leg to detect whether, and if so to what extent, there has been a thickening of the intima of the vessel.

What is the relationship between these practices? There is, says Mol, a textbook explanation. It says that arterial disease leads to a thickening of the vessel intima. Beyond a certain point this leads in turn to a fall in blood pressure and, again beyond a certain point, this begins to interfere with the blood flow in the legs. When this happens, the leg muscles don't receive enough oxygen during exercise—and the result is claudication, pain, upon walking.

This textbook story is realist in character. It assumes that there is an object—lower-limb arteriosclerosis—out there that manifests itself in various ways. If one looks as what goes on in hospitals and doctors' surgeries there are moments, indeed, when this story is quite unproblematic. Thus a patient may arrive complaining of claudication, and when his blood flow is measured, it turns out to be inadequate. He may then be treated in one way or another. Rarely—and this is only after all possibilities of treatment have been exhausted—there may be need for amputation. If this happens, the amputated limb is sent to the pathology laboratory and the sectioning of the vessels will reveal, if all goes as anticipated, substantial thickening of the vessel intima.

So there are occasions when it is possible to say that there is an object out there, "lower-limb arteriosclerosis," that manifests itself in a series of different ways. It often turns out, however, that the three medical practices described above come to different conclusions. A patient may be suffering from claudication, but there appears to be no pressure loss at the ankle. Or a patient whose blood vessels seem to be occluded turns out to have no pain on walking. There are endless case conferences in the hospital dealing with problems such as these. There are also many strategies for explaining these inconsistencies away. And, in particular, there are ways of dealing with the pressing question as to what should be done for a patient who is in pain or whose lifestyle has been restricted.

Annemarie Mol is a philosopher who argues that the three practices generate three different and sometimes very badly coordinated arterioscleroses. In the plural. And this is the problem of difference: practices may and often do generate multiple rather than singular objects.

TSR2 brochure, or so I want to suggest, embodies and performs a number of these.

Syntax

Let's start very humbly, even though the point is obvious and may be made very briefly. The brochure coordinates different objects, different TSR2s, in part by means of *syntax*. Exhibit 2.2 is an example of this: TSR2 is a weapons system, or so we learn; it is capable of reconnaissance roles; it is capable of nuclear roles; it is capable of high-explosive strike roles; it is capable of operating in all weathers; and it is capable of doing so with a minimum of ground support facilities. How then is TSR2 made into a single object, rather than four or five? Of course the answer is by syntax. What might be several aircraft are being grammatically turned into one.[4]

Physical Structure

The second strategy is almost equally humble. This is the physical structure, not of the aircraft, but of the brochure. As I've said, the latter is sixty-odd pages long. More important, these pages are numbered and bound, together with a cover, a title page, and a table of contents. I shall have more to say about the table of contents shortly. For the moment let me just observe that the cover (exhibit 2.7) may be understood as a mechanism that glosses whatever it is that will follow within the pages of the brochure. Announcing itself in bold type, "TSR2," and then adding in a smaller cursive typeface, "Weapons System," the cover frames or coordinates the contents of the brochure. Presumptively, then, and as a result of this, everything within the brochure—and that includes the above exhibits—will have "something to do" with the TSR2. Physical structure, then, is a second strategy for coordinating disparate objects or object positions.

Tabular Hierarchy

Look now at exhibit 2.8, which reproduces the table of contents of the brochure. This makes more links, coordinates further versions of the TSR2. But how does it produce its coordinating effects?

Michel Foucault offers us the classic response. A table constitutes and juxtaposes components in a two-dimensional array. It generates new forms of visibility, new visual relations, which means that it cre-

EXHIBIT 2.7
Brochure Cover
(British Aircraft
Corporation 1962;
© Brooklands
Museum)

Contents

Page	'ENGINEERING' SECTION
30 – 31	FUSELAGE LAYOUT
32 – 35	AIRCRAFT STRUCTURE
	Fuselage – Wing – Empennage – Landing Gear

Page	
4 – 5	INTRODUCTION
6 – 7	LEADING PARTICULARS and 3-VIEW GENERAL ARRANGEMENT

	'PERFORMANCE' SECTION
8 – 9	AIRCRAFT PERFORMANCE
	Speed – Take-off – Landing – Ferry Range
10 – 11	MISSION RADII
12 – 13	ATTACK SYSTEM – NUCLEAR ROLE
14 – 15	ATTACK SYSTEM – H.E. ROLE
16 – 17	RECONNAISSANCE SYSTEM
	Photographic Reconnaissance – Survey Facility
18 – 19	RECONNAISSANCE SYSTEM
	Sideways Looking Reconnaissance Radar – Linescan

	'OPERATIONS' SECTION
20 – 22	AIRFIELD OPERATION
23	STRATEGIC DEPLOYMENT
24 – 25	RADII OF ACTION
26 – 27	NAVIGATION SYSTEM
28	TERRAIN FOLLOWING
29	COMMUNICATIONS

36 – 37	POWER PLANT
	Fire Protection – Cooling and Ventilation – Auxiliary Power
38 – 39	CREW STATIONS
40 – 41	FUEL SYSTEM
42 – 43	HYDRAULIC SYSTEM
44 – 45	PRESSURIZATION AND AIR CONDITIONING
46	ANTI-ICING
46	OXYGEN SYSTEM
47	ELECTRICAL POWER and LIGHTING
48 – 49	FLYING CONTROLS
50 – 51	NAVIGATION AND ATTACK INSTALLATION
52 – 53	COMMUNICATIONS INSTALLATION
54 – 55	ARMAMENT INSTALLATION
56 – 57	RECONNAISSANCE INSTALLATION
58 – 59	GROUND EQUIPMENT
	General Servicing Vehicle
60-61	GROUND EQUIPMENT
	Universal Lifting Trolley
	Hydraulic Test Equipment
	System Test Equipment

EXHIBIT 2.8
Contents (British
Aircraft Corporation
1962, 3;
© Brooklands
Museum)

ates a subject position that escapes the linearity and the syntax of text
by virtue of its construction as an overall vantage point. This means,
to use another metaphor, that a table draws things together. It draws
them together by juxtaposing them spatially and by assuming, in that
juxtaposition and ordering, that they are in some way linked. It thus
performs the assumption that the different elements listed are simi-

lar in kind. And in the present instance, if one makes the link back to the framing of the brochure, that they concern the TSR2 in one way or another.

So the table implies and performs a form of coordination. When nouns and the different specific object positions appear in a list or a table, they are being made to go together. But this is simply a first step. For in the present instance at least, these relations of visual simultaneity also perform *relations of hierarchy.* Components of the table, its elements, are being coordinated in ways that assure their asymmetry. Thus, it is not simply that what become the more important features of the object tend to come in the earlier pages of the brochure (though this is certainly one of the effects being achieved, both in the table of contents and in the overall structure of the publication). It is also that the elements in the table are ranked into three different levels. There are the three main sections, on "Performance," "Operations," and "Engineering"; a number of subsections, such as "Aircraft Performance," "Radii of Action," and "Ground Equipment"; and then there are subsections to those subsections, which have to do with objects such as "fire protection" or "system test equipment."

So the table of contents, by virtue of its visible organization, not only homogenizes, not only proposes that everything in the table somehow or other goes together, but also makes a hierarchy in three levels. This means that the table of contents is like an organizational chart or an arborescence:[5] the various elements are being defined, performed, and indeed guaranteed. Smaller parts are being "included" in the larger sections. They become specific "aspects" of larger unities—and, no doubt, of the TSR2 aircraft as a whole. Thus the reader is readily able to see that what has now become "the top" depends on, or is composed by, the links between a series of more specific components that have their role to play in the system "as a whole."

In short, the table is a third strategy for coordinating disparate objects and relating them together to form a unity. And the particular alchemy of the table is the way in which it returns, constantly, to that which is made central—ultimately, though here implicitly, the TSR2—and performs what, echoing Jacques Derrida's concern with necessary slippage (1978), one might think of as *strategic deferral,* for that which is left out is performed as "detail," as "technical detail."

Note that the logic of strategic deferral—and the way in which it produces subject and object centering—extends beyond the covers of the brochure. For if this is only sixty pages long, then much must be excluded from its pages and deferred because it is "less important" or "relevant" than what is mentioned. Nevertheless (and here is the assumption of this strategy for coordinating, which is also a guarantee), it *might* be unpacked if the curious chose to look at "supplementary documentation" in some "technical manual" for reasons that are made to be good because, for a complex object such as an aircraft, it is important to perform links that are many layers down the hierarchical-technical system.

To summarize: the table, its structure, and its deferrals produce a hierarchy that generates a subject that has focus but also a coordinated object, one that hangs together because it has been constituted as a set of hierarchically related parts or aspects that combine to produce a unitary whole.[6]

Perspective

Earlier I suggested exhibit 2.1 is perspectival in character. But so too are exhibits 2.9 and 2.10. It is clear to all but the naive reader that these are pictures of the same object. In other words, we are justified in detecting the operation of yet another strategy for coordinating possibly different objects—that of perspectivalism.

EXHIBIT 2.9
Dispersed Airfield
Refuelling (British
Aircraft Corporation
1962, 40;
© Brooklands
Museum)

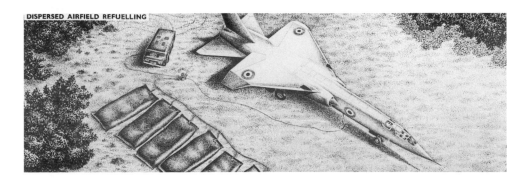

DISPERSED AIRFIELD REFUELLING

I shall have more to say about this later, so let me just note for the moment that perspectivalism assumes a world that is Euclidean in character; that is, it assumes that the world is built as a three-dimensional volume occupied by objects. These objects, which in-

EXHIBIT 2.10
Front View (British
Aircraft Corporation
1962, 2;
© Brooklands
Museum)

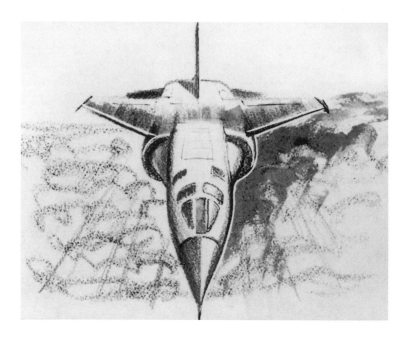

clude a variety of positions for viewing, have locations and (at any rate in the case of objects) themselves occupy three-dimensional volumes.

This is why when we look at individual perspectival drawings of the kind in exhibits 2.9 and 2.10, we tend to see a three-dimensional object, for instance an aircraft, rather than some lines on a sheet of paper. The theory of linear perspective holds that we project the lines that appear on the paper in such a drawing back into a Euclidean volume of space. That volume is occupied by a three-dimensional object that would, had it been located in such a space, have traced itself onto a two-dimensional surface in a way that corresponds with the lines on the sheet of paper. Thus in linear perspective a viewer or subject position is made that "sees" an "object," the aircraft, even though it sees only a sheet of paper. Or, to put it differently, what is on the sheet of paper tends to produce the sense of an object because it helps to *reproduce* a Euclidean version of reality.

Furthermore, and an important part of this strategy, different perspectival sketches are easily coordinated within the system. There is a formal projective geometry for saying this, but let me put it informally. One of the Euclidean assumptions of perspectivalism is that a single three-dimensional object can generate multiple two-dimensional perspectival depictions. Objects, the same objects, simply look different if we look at them from different standpoints. And this is what is happening here. The coordinating assumption is that there is an aircraft, the TSR2, and that it is fixed in shape. That singular fixity will generate all sorts of possible two-dimensional perspectival configurations. So long as the depictions conform to these configurations and do not demand an impossible three-dimensional object, then we tend to see the same three-dimensional object when we look at different perspectival drawings, as we do here.

Cartography

Perspectivalism is only one projective strategy for visual coordination; there are many more. For instance, art historian Svetlana Alpers writes: "the Ptolemaic grid, indeed cartographic grids in general, must be distinguished from, not confused with, the perspectival grid. The projection is, one might say, viewed from nowhere. Nor is it to be looked through. It assumes a flat working surface" (Alpers 1989, 138).

Thus cartography is another strategy—or better, a series of strategies—for coordinating disparate specificities.[7] We have already come across one of these in exhibit 2.6. Exhibit 2.11 is somewhat similar. Both are maps drawn, like all maps, to a particular projective con-

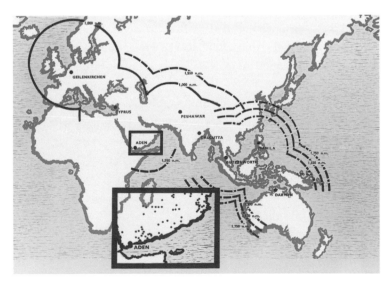

EXHIBIT 2.11
Operation (British
Aircraft Corporation
1962, 24;
© Brooklands
Museum)

vention that (at any rate here) "flattens" a world which (as with per-spectivalism) is taken to occupy a three-dimensional volume. Spe-cifically, it unwraps what is taken to be the surface of a spheroid (in the case of exhibit 2.11, a part of that surface) and to flatten it onto a two-dimensional surface. In doing this, it locates, juxtaposes, and interrelates geographical features to generate what, as Alpers notes, is a view from nowhere—nowhere, that is, in the kind of Euclidean perspectival space generated in exhibit 2.1. This is because the eye (and the projection as a whole) is located outside Euclidean space, even though it is generated by transforming that space.

The view from nowhere is thus made in a way that sees things that could never be seen within perspectivalism. Or, to put it a little differ-ently, it makes a centered viewpoint, a centered subject, using a flat-tened working surface that coordinates objects taken to be out there. It is like the table except that the relations performed by the two work-ing surfaces, the contents and the map, are different.[8] In the former case we were dealing with objects that were being related together into a hierarchy, whereas here we are dealing with the performance of spatial relations.

But we're interested in the aircraft. So where is the TSR2 in these projections? The answer is that it is located on the working surface of the map—but also that it is invisible. Quite simply, if it were de-picted in terms of the scaling conventions used in these projections, it would be submicroscopic in size. So the aircraft is there: it is as-sumed that it is indeed located on the surface of the map, which is also the surface of the globe. But because we cannot see it, we need to mobilize further conventions or strategies if the maps are to do useful coordinating work.

Let's say first that the two maps are multiply connected. As I have indicated above, they represent the operation of similar cartographic conventions. Second, they appear in the brochure, so for physical rea-sons they both presumptively have to do with the TSR2. Third, that presumption is strengthened by the fact that they are bound together on facing pages. But we need more than this. In particular, we need to make the TSR2 visible. So how does this work? The answer is that the two maps mobilize different conventions.

Exhibit 2.6 works because there is an understanding that mobile objects traversing geographical space may leave huge cartographic traces in their wake, traces that here take the form of thick lines and arrows. These traces disrupt the scaling conventions, being in those terms several hundred kilometers wide. However, this disparity is no problem for the informed reader. This combination of conventions, which applies just as well to the movement of buses in a public transport system, makes it possible for the viewer from nowhere to "see" movement on a cartographic surface. Specifically, what the viewer sees or learns here is that the TSR2 is a global traveler. Or, to put it differently, that the same object may move around and be found in the United Kingdom and Australia.

Exhibit 2.11 undoes the invisibility of the aircraft in another way. Again the surface of the map is covered with lines that must, in terms of cartographic understanding, be fifty kilometers wide. However, this time convention tells us that these have nothing to do with imaginary traces left behind by flying aircraft. Instead they represent the boundaries of areas—areas, as is obvious, that may be overflown by the TSR2 in its sorties if it is based at one or other of the locations named on the map.

In all this we are unearthing a series of cartographic and cartographically relevant strategies for depicting the geographically relevant attributes of objects. But we are also learning something more about the ways in which these intersect and coordinate with one another to produce a singular object with particular properties. Thus, though the naive reader was denied this knowledge, I started this essay by noting that the brochure was aimed, perhaps in particular, at the senior members of the Royal Australian Air Force. Now it becomes clear that in their juxtaposition and their mobilization of several different cartographically relevant conventions, these maps bring together two features of the TSR2 of great potential importance to Australian strategists: first, its ferry range, and second, its operational range. The aircraft that can fly round the world is coordinated with the aircraft that can undertake very long-range missions into communist China. The triangulation between the conventions of cartographic projection, the traces left by moving objects, and the depictions of areas interact to ensure that we are here dealing with one and the same machine.

System

Exhibit 2.12 takes us to the navigation system and to an example of another strategy for coordinating object positions: discursive (and, as we shall shortly see, pictorial) monitoring and self-correction.

How does this work? The strategy coordinates object positions in a way that suggests monitoring and self-correction will put each other right, indeed perhaps rebuild one another, should the links between them or the positions concerned start to weaken. Such is the point of the term "correction" in exhibit 2.12 and the rationale for talking of a "weapons system" in exhibits 2.2 and 2.13.

These terms and this strategy are both less hierarchical than those built in the table of contents. On the other hand, they echo a substantial literature on large technical systems in technoscience studies, a literature that lays stress on the interconnected and interlocking character of technical innovation.[9] The argument is that large technical systems—and here TSR2 is being treated as an exemplar—look at and talk to themselves reflexively.[10] That is, they bring themselves into being and sustain themselves because they build, or take the form of, feedback systems.

Exhibit 2.13 generates components—let's say object positions— that are coordinated to perform and stabilize the "complete TSR2 weapons system." "In-built test facilities, pre-checked packages for armament, etc.," such are specific objects that work together in this discursive strategy to secure the stability and continuity of the TSR2 object through a series of different positions.

The strategy takes different forms in different places. For instance, exhibit 2.13 works by colonizing alternatives, by simply obliterating them, or by rendering them irrelevant. This is the point of the talk

---

EXHIBIT 2.12 "Fixing consists of comparing the computed position of the fix point with the actual position of the point as shown by radar. Both these positions are shown on the navigator's radar display and his action in comparing these produces a signal proportional to the displacement between them. This signal is fed to the digital computer where it is used to correct the computed dead reckoning position and may be used to feed an azimuth correction to the inertia platform if necessary." (British Aircraft Corporation 1962, 26)

EXHIBIT 2.13 "The complete T.S.R.2 weapons system is designed for mobility and flexibility in operation, reversing previous trends towards reliance on major base facilities. It can be deployed rapidly throughout the world with nominal support and is then ready for immediate operational use to an extent depending on the level of support. In-built test facilities, pre-checked packages for armament, etc., and an auxiliary power plant for operating aircraft electrics, cooling and other systems, are used during the turn round, thereby avoiding reliance upon complex support equipment. All support equipment is air-transportable." (British Aircraft Corporation 1962, 5)

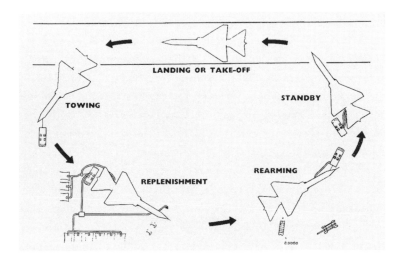

EXHIBIT 2.14 Turn Round at Dispersed or Primitive Airstrips (British Aircraft Corporation 1962, 20; © Brooklands Museum)

of rapid deployment, in-built tests, and pre-checked packages. If airfields are unreliable and cannot supply support-and-test equipment, then these too can be re-created as a part of the system.[11] This means that what is being constituted as a single aircraft, like the experiments described by Karin Knorr-Cetina in high-energy physics,[12] ends up interacting with itself rather than with the outside world. In this way a number of different aircraft—the aircraft at a fully equipped airfield, but also an aircraft at a primitive airstrip somewhere in the forests of Germany—are coordinated: they are made substantively, but also functionally, coherent. We are dealing, in all senses, with "the same" aircraft.

The looplike and self-sustaining character of this strategy of co-ordination is visible in exhibit 2.14. This (or so the caption tells us) is to be understood as a diagram of "turn round at dispersed or primitive airstrips." In this depiction the loop starts (and ends) with landing and take-off, moves through icons that depict towing, replenishment, rearming, and standby, using (as the text observes) "self-contained facilities [which] can be used for normal operations." Many coordinating conventions are being deployed here, textual and iconographic. But in the present context it is the arrows that are most significant. The conventions for reading these generate a viewer who is not naive but understands that the five icons for the aircraft stand not for five different aircraft but rather for one: a singular aircraft that is being displaced through time, and perhaps (though this is less clear) through space. A singular aircraft is being made that will be returned to the sky even though it is far removed from major base facilities.

A version of the same systems singularity is also deployed in one of the maps discussed earlier—the ferry-range map of exhibit 2.6. We have discussed the cartographic practices mobilized here, and also the conventions for tracing the movement of small objects onto mapped surfaces. But the latter with its lines and arrows also performs a further version of systems coordination. As in exhibit 2.14, this takes the form of lines and arrows which go round, in loops. The particular rhetoric of TSR2 singularity here thus not only performs an aircraft that can fly long distances once it is fueled up, but also performs a TSR2 that can go out, for instance from the UK, but that *can also come back.* This, it should be added, is not the trivial matter that it might seem, given prevailing headwinds and the need to plan for adverse conditions when seeking to land on tiny islands in the middle of the ocean.

Each of the exhibits I've touched on in this section contains loops. Each performs loops. And the strategy for coordination depends on the successful manufacture of loops. For, in a systems world, the world of cybernetic self-maintenance, properly built loops are reassuring. They correct themselves. They secure an environment in which coherences may sustain themselves and that does not distort what is passed round the loop. So it is that such loops, or the connections that afford such loops, generate what Bruno Latour (1987) calls immutable mobiles. Objects remain "the same" even as they move and displace themselves.

Speed/Heroism

As a final example of the working of the strategies of coordination in the brochure, I want to touch briefly on the issue of speed.

Like exhibit 2.2, exhibit 2.15 deploys syntactical and discursive conventions to create an object that is capable of flying both fast and low. Exhibit 2.16 uses graphing conventions that are somewhat related to those of cartography, both to identify an aircraft that is capable of the long-range missions identified by more direct cartographic means in exhibit 2.11 and again to offer a message about speed. But the making of an object that is singularly fast uses many more conventions, and some of them are much less direct in character. For instance, exhibit 2.1 depicts an aircraft (which we may now agree is the TSR2) from behind and below. Though this is not given in its perspectivalism, a competent reader will also note the undercarriage is retracted. This means that in the depiction the aircraft is being made to fly, made to move, though it is true that we are given no clues as to how fast it might be moving. But this is not the case for the front cover (exhibit 2.7). Like exhibit 2.1 this is again in part produced by the technologies of perspectivalism. At first sight it might seem that the viewpoint is that of the pilot. But this isn't quite right because the pilot, confined to his cockpit behind the heavy canopy that protects him, would not enjoy a spectacular all-round view of the kind on offer here. In which case the representation may not so much be what the pilot sees but rather what the aircraft itself can see. Perhaps, then, it is a representation of the view the aircraft would enjoy as it flew at two hundred feet.

I have mentioned perspective, but there is another visual strategy at work here, one that is crucial for performing a distinction between

---

EXHIBIT 2.15 "TSR2 is designed to operate at 200 ft. above ground level with automatic terrain following, at speeds of up to Mach 1.1. It is capable of Mach 2 plus at medium altitudes." (British Aircraft Corporation 1962, 4)

EXHIBIT 2.16
Sorties (British
Aircraft Corporation
1962, 10;
© Brooklands
Museum)

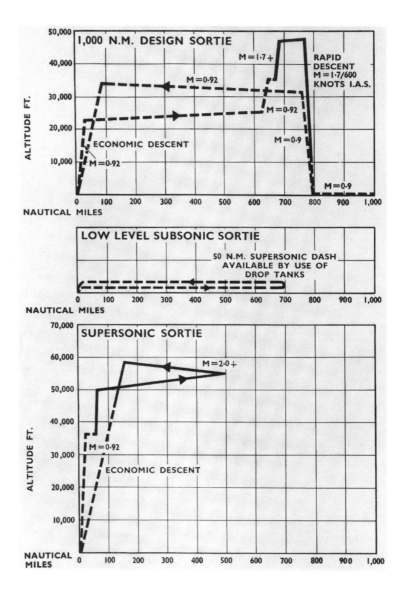

stasis and movement—and which also tells us the direction of that movement. It works, as is obvious, by blurring many of the lines and surfaces in the visual depiction. The principle is straightforward: as in a cartoon strip all those lines not parallel to the direction of travel are blurred. If the viewpoint itself is traveling, then static objects are blurred. Seen from a static position it is, of course, the other way round. However, it is the first of these alternatives that is being mobilized here. And since the convention is indeed superimposed on a version of perspectivalism, those lines that are not blurred converge to a vanishing point: the place into which the aircraft will shortly disappear.

As is obvious, there is a connection between all of these exhibits. All embody strategies for *making speed,* albeit by different means. Through these different texts and depictions the TSR2 is being turned into a very fast aircraft. But in exhibit 2.7 something else is going on too. Here speed is being made a relative matter, turned into a question of contrast. A division between movement and statis is being performed. This is implicit in the technology of blurring superimposed onto perspectivalism. This exhibit suggests that the aircraft is only present for a split second. Right now it is, to be sure, above the buildings that are dimly discernible below. But in half a second they will be gone. They will have disappeared as the aircraft itself disappears into the vanishing point.

And what is the significance of this? I offer the following suggestion: we are witnessing not only speed but also a depiction that helps to reflect and perform a particular version of male agency. Thus I take it that the front cover is telling, in a way that the textual and graphing conventions of exhibits 2.15 and 2.16 do not, that this is not just a fast aircraft and one that flies low. We are also being told that it is an exciting aircraft to fly. That it is a thrill to fly. That it is, in short, *a pilot's aircraft.*

The creation of speed is not, to be sure, itself a strategy of coordination. Rather, it is an effect of a series of such strategies. It is crucial, however, for all sorts of reasons. Some of these are "technical" in character (to use a term of contrast that I will try to undermine in chapter 6). There are, indeed, technical or strategic reasons for the aircraft to fly very fast and very low. But others are not. Thus I suggest that speed is the raw material on which another effect builds itself: the depiction and performance of heroic agency. In this way it helps to make a particular kind of reader—not simply a "technical" subject who responds to the technical attributes of the TSR2, one who wants to know how far it can fly, whether it can defend the Australian Northern Territory from the Indonesians or the communist Chinese, or whether it can fly under the enemy radar screen. It also helps (and here is the other half of that dangerous contrast) to make a subject that is aesthetic, indeed erotic, one that enjoys fast and even dangerous flying. Thrills and spills: there are coordinating strategies for linking these together. And unless we can generate a reader of this kind from the specificities of the materials of the brochure, then we are, indeed, missing out on something very important about the effects of the strategies for coordinating those different object specificities.[13]

Decentering the Object

The naive reader does not exist, except perhaps as a methodological fiction. But the creation of the naive reader throws the problem of difference into relief. This is because it generates many objects or object positions and many subjects or subject positions. It brings a flock of aircraft into being, together with a library full of different and disparate readers. And so it generates an inquiry, the inquiry into coordination, the inquiry into how the various subject and object positions are aligned with one another, and the inquiry into the *strategies* for such coordination. The inquiry, then, is into how singular subjects and singular objects are made.

But wait a moment. Now the alarm bells start to ring. Ever since Lacan (or is it Freud?) there have been questions and doubts about the centered subject. Ids, egos, superegos, and their endless descendants, we have become habituated to the idea that the self is divided, the subject a set of more or less unsatisfactorily related subject positions. So the idea of the decentered subject is scarcely new—though, to be sure, it has taken on new life in recent work in cultural studies, where the possibility that noncoherence between different subject positions might also be desirable has taken root.[14] But if the idea of the decentered subject is not new, then what of the decentered object? What of the object that does not hang together? Or holds together only partially?

Here the arguments have not been properly made or explored. But such is the prospect that we face if we take the problem of difference seriously. And it is the problem that we all face if, as I have in this chapter, we start to wash away the *assumption* of singularity, the *presupposition* that, whatever we might study and whatever we might interact with is indeed a single, coherent, and centered object that is out there. A single object that we may come to know in this way or in that. A single object over which we may have different perspectives. But, nonetheless, a single object.

So that is the abstract version of the story. But what if we return to the TSR2 and ask, was this a single object? Was it *an* aircraft?

The answer to this question is, at least in part, an empirical matter.[15] It is conceivable that the strategies for coordinating the various TSR2s, for making them singular, indeed dovetailed together to generate a unity. But I doubt it. I doubt it very much. And this is not primarily because the TSR2 project encountered a series of difficulties that became the topic of endless debate in the policy and procurement literature. It is rather because, once we look at things in this plural way, any singular object immediately becomes an effect—and a more or less precarious effect. Yes, arteriosclerosis. Yes, alcoholic liver disease. Yes, a water pump. Yes, a program of medical screening or health advice. Yes, a pregnancy.[16] And yes, an aircraft. All of these are more or less singular but *also* more or less plural. And if the well-publicized difficulties of the TSR2 project are relevant here, it is simply because they make it easier for us to see some of the non-coherences.

For this business of multiplicity and coordination is not a clever game dreamed up by poststructuralist philosophers or students of postmodern social science. Or if it is a game, then it is one that is also real enough. Indeed it is one that is deadly serious. Exhibit 2.17 tells of the inability to coordinate the development of the subsystems of the aircraft to produce a desired coherence in the form of a single and coherent object. It thereby discursively undoes the work of coherence performed by many of the earlier exhibits.[17]

EXHIBIT 2.17 ". . . it proved intrinsically impossible to co-ordinate the airframe, electronics and engine work." (Williams, Gregory, and Simpson 1969, 53)

Exhibit 2.18, posed in the language of policy, tells of the inability to make a single aircraft in "reality" that would fit the "concept" of such a weapons system.[18] This too undoes the coordinating work of

EXHIBIT 2.18 "The TSR-2 weapon system was an extremely advanced concept, combining several roles in one aircraft, attempting to achieve compatibility in performance which had not previously been attempted, and projecting air power requirements well into the '70s. Here, perhaps, is the basic weakness of the TSR-2 concept, the attempt to meet too many new and complex specifications at the same time." (Williams, Gregory, and Simpson 1969, 20–21)

the brochure and in particular that of exhibit 2.2, which insists on the necessary integration of a single weapons system. In exhibit 2.19 we learn the need for separate battlefield and deep-strike aircraft: the expense of the latter made it impossible to imagine that it could ever be the same as the former. Again, then, this is a performance of disaggregation. While in exhibits 2.20 and 2.21 we learn that in this version of similarity and difference, deep-strike aircraft could never be conventional: the idea that a deep-strike aircraft would be both conventional

Perspectives, Epistemologies, and Ontologies

Linear perspective, and indeed its alternatives, works on the assumption that it is possible to distinguish in a three-way division between a viewer, that which is viewed, and a representation of that which is viewed.[21] It assumes that these are distinct, and that the problem facing the artist is technical in character: that of translating that which is into that which represents it. This has various consequences:

—It models the subject as a single point location within or to one side of a three-dimensional geometrical space.

—In one way or another, it accordingly solves an *epistemological* problem: the problem, that is, of providing warrantable or workable knowledge of the world.

—It is (tautologically) perspectival in character. That is, in its enactment it implies a *reduction* of the world that might be seen from many viewpoints to what may be depicted from a single viewpoint. There is always the possibility of other reductions from different viewpoints. Representation never exhausts the possibilities. There are always others.

—It rests upon and performs a family of related *ontologies:* that is, assumptions about the nature of existence or being, about what there is. In particular, it assumes that there is a more or less stable world "out there" that may be depicted from one perspective or another.

There is a relation between epistemology and ontology here. An ontological assumption is *performed* in tackling an epistemological question. Or, to put it in a more pointed manner, the possibility that an ontology is being created or performed is concealed by the focus of attention on epistemology.

Distinctions between perspectivalism and an approach based on semiotics such as that used here, include the following:

—Perspectivalism trades in epistemology and effaces ontology, whereas semiotics trades in ontology: it is a method for exploring the simultaneous creation of objects and subjects.

—Perspectivalism describes what is. Semiotics tells about the making as well as the knowing of things.

—Perspectivalism solves the problem of multiplicity or difference by reconciling or explaining different views or perspectives because it says that they are looking at a single object from several points of view. Semiotics says that different objects are being produced, and then asks how, if at all, they are connected together in order to create a single object.

and nuclear was, in effect, not just a noncoherence but a contradiction.[19]

EXHIBIT 2.19 "Logically what was needed were two weapons systems, one for carrying out, economically, conventional operations in the battlefield areas, and a second for deep penetration nuclear strike operations. . . TSR-2 was too expensive to risk in th[e former] role in anything other than very bad weather." (Williams, Gregory, and Simpson 1969, 42)

To be sure, these exhibits take the form of belated wisdom. It is easy to be wise after the event. But that is not the point of citing them. It is not that they are right and the brochure is wrong. I have no desire to take sides. It is rather to show that the work of object coordination and object disaggregation goes on—and on. It is to suggest that the singularity of an object is precarious, uncertain, and revisable.[20] And thus it is to suggest that the issue of what there is and what there could be, whether the objects in the world are centered or decentered, singular or multiple, whether they are both, or whether somehow or other they are fractional, this is not simply a question of playing postmodern games. For if we start with a naive reader, this is not to celebrate naïveté but rather to lead us to questions of similarity and difference. And these are questions that are real enough. They have to do with

---

EXHIBIT 2.20 "Another point that worried me [about TSR2] was that a supersonic aircraft was not likely to be used for close-support of troops fighting on the ground. Were a war to erupt on the European mainland, I could hardly imagine that the Russians would wait to find out whether aircraft making deep strikes on targets within their territory were carrying conventional bombs, leaflets or nuclear weapons. If we and the Russians meant what we were saying, the response would most likely be nuclear." (Zuckerman 1988, 214–15)

EXHIBIT 2.21 "But above all I could not see any strategic sense in the notion that the TSR2 could be operated as a fighter-bomber armed with nuclear bombs for use on a European battlefield. The idea of nuclear field-war was nonsense." (Zuckerman 1988, 215)

---

coordination. They have to do with the strategies that secure coordination and the ways in which such strategies intersect to build up or
break down similarities and differences. They have to do with what
there is, and what, in a fractional world of coherence and noncoherence, there might be. They have, in short, to do with ontology.

The distinction between the public and the private is a distinction internal to bourgeois law. — Louis Althusser, *Ideology and Ideological State Apparatuses*

Critical theory is not finally about reflexivity, except as a means to defuse the bombs of the established disorder and its self-invisible subjects and categories. — Donna Haraway, "A Game of Cat's Cradle: Science Studies, Feminist Theory, Cultural Studies"

Here is one beginning.

Perhaps, then, there are many objects and many subjects. This is the origin of the problem of difference, the problem of multiplicity. If it sometimes appears that there are singular objects and singular subjects then, somehow or other, object positions and subject positions have overlapped and been linked. This is the problem of coordination, the problem of coherence. Multiplicity and coordination: the two come together. But how do subject positions cohere? How are knowing subject positions constituted and chained? And what happens if it doesn't work? What happens if we end up with broken subjects?

Here is another.

I've been puzzling for some time about this: the problem of the public and the private, or the role of the personal in ethnography or history. Let's put "the personal" into quotes. I've been puzzling for some time about the problem of "the personal" in writing in science and in social-science writing: how it works; what it does. My puzzle refracts itself in my own writing on technoscience. The question is whether I should rigorously try to keep the "personal" out. Such would surely be the dominant response. But supposing I were to let it in, then what should I do about it? How might I handle it? What kind of job should it be doing there? These are the issues that I investigate in this chapter.

I have learned much about stories from the anthropologist of science, Sharon Traweek (1988a, 1992, 1995a, 1995b, 1999). One of the things I have learned is that when we tell stories — including those that do not appear to come to the point — they are performative. So there are two points here, one to do with performativity, and the other to do with what we might think of as "indirection," that is, the absence of a visible focus, a place within the story that says in as many words what it is "really about."

On the question of performativity the argument is quite simple. As I suggested in the introduction, stories are performative because they also make a difference, or at any rate might make a difference or hope to make a difference. The question of indirection is trickier. Like Sharon Traweek, however, I am committed to indirection. I want to imagine alternative versions of what it is to theorize; versions that avoid the hierarchical distributions between theory and data, or theory and practice; versions that instead perform multiplicities and interferences; versions that come to terms, in the way they perform themselves, with the postmodern possibility that it is not possible to draw everything together into a simple and singular account; versions of theorizing that, in other words, are allegorical rather than literal in form.[1]

I will return to both performativity and indirection in later chapters. Here, however, I introduce the terms explicitly—yes, in as many words—because this chapter is composed of performative but largely indirected stories that have to do with the "personal." I do this because I want to make a difference in the way we imagine what we now think of as the "personal," the "analytical," and indeed the "political." I think that if we do it right, it turns out that the "personal" is not really personal any longer.[2] Instead it is an analytical and political tool, one among many that might allow us to defuse some of the bombs of the established disorder.

1965

I will start with a story about politics and an aircraft, an aircraft as seen by a young man. The young man was called John Law. But the past is at least in part a foreign country, and because they do things differently there I will recount it in the third person.

The air was heady. A senile Conservative government had been defeated at the polls. It was a pity that it hadn't been overturned by a larger margin. But the country had a Labour government, a government that was going to undo the harm done by "thirteen wasted years" of Tory rule. It was going to abolish medical prescription charges, renationalize the steel industry, and (most important in the present context) cut out waste on "Tory prestige projects." Such was the promise.

On election night one of his lecturers told that young man in an all-night café for transport drivers and railwaymen in the center of Cardiff, "We've got the bastards now." And that is what he believed.

That was in October 1964. Seven months and a number of disappointments later there was an announcement: the government was going to cancel one of the much-hated "Tory prestige projects," a military aircraft called the TSR2. I don't think the young man knew very much about the TSR2, but he knew some things. Perhaps he knew, or at least sensed, three things.

First, he believed that this project was a monstrous waste of money, that it was vastly over budget and that it was behind schedule. Such, at any rate, was what the government said, and he had no particular reason to doubt that it was true.

Second, he was told that this aircraft was a part of Britain's "independent nuclear deterrent." This was in itself a reason for canceling it because he was a supporter of CND, the Campaign for Nuclear Disarmament. This was a largely left-wing pressure group that wished Britain to give up its nuclear weapons unilaterally. Again, he didn't know too much about the detail of these arguments, but he knew what he thought. And he also knew that the new Labour government, in its first seven months, hadn't canceled its Polaris submarines. Indeed, it had reaffirmed the importance of this central part of the British nuclear force and had scrapped only one of the projected submarines.[3] This was one of the larger disappointments. So, like other CND supporters he'd felt betrayed by a Labour Party that had toyed with unilateralism. This meant that the TSR2, though small beer by comparison, was at least a gesture in the right direction.

Third, when he learned of the cancellation, at the same time he also felt a sense of disappointment. But why? The specifics of this dissatisfaction are, shall we say, a little obscure. I hope that some of them will become clearer in what follows. But for the moment let's just observe that he'd seen pictures of this aircraft on television and in the newspapers. It was in the early stages of its flight-proving program and the manufacturers had released film footage of it. There it was, taking off, flying around, and landing. And, though I don't think he said this to anyone, the aircraft appealed to him. It appealed to him how? Let's say that it appealed aesthetically as powerful, masterful, sleek. To witness it in flight was obscurely or not so obscurely thrilling.

The gendering tropes are obvious enough: the business of men and their machines, control, force, and power.[4] But the fact that they are clichéd makes them no less real. So, though I don't think that the cancellation of the TSR2 was that big a deal one way or the other, he was nevertheless somewhat ambivalent when he heard the news.

Situated Knowledges

Donna Haraway has written one of the more influential papers in recent feminist writing on technoscience. It's on situated knowledges. It does many things, this paper, but one of them is to investigate the optics of knowing, an optics performed in the natural — and no doubt in many of the social — sciences. This optics seeks to perform itself as disembodied, as removed from the body, indeed as having nothing to do with the body: "The eyes have been used," writes Haraway, "to signify a perverse capacity — honed to perfection in the history of science tied to militarism, capitalism, colonialism and male supremacy — to distance the knowing subject from everybody and everything in the interests of unfettered power" (Haraway 1991d, 189).

Vision has been disembodied in what she calls the god-trick, "the god-trick of seeing everything from nowhere" (Haraway 1991d, 189). But, she says, vision is never from nowhere. I talked about this in the last chapter: vision is always from somewhere, even if that somewhere takes the form of a cartography that projects itself from nowhere Euclidean in particular. To put it a little differently, vision always embodies specific optics, optics that vary from place to place and, for that matter, from species to species. Which suggests that (1) any reflective — or even pragmatic — optics that claims to stand back and see it all from a distance is a form of mythology; (2) to the extent it is built into a particular mammalian visual system, such an optics is in any case one that is highly specific; and (3) the notion, an alternative notion, of objectivity may be rescued if the body is, as it were, put back into the process of seeing. Donna Haraway again: "objectivity turns out to be about particular and specific embodiment, and definitely not about the false vision promising transcendence of all limits and responsibility. The moral is simple: only partial perspective promises objective vision. This is an objective vision that initiates, rather than closes off, the problem of responsibility for the generativity of all visual practices" (Haraway 1991d, 190).

There are various components in this turnaround, this attempt to recolonize the notion of "objectivity" for something that is local and situated. Let's mention just two. One is the recommendation—certainly not Haraway's alone—that a commitment to specificity implies a willingness to accept a kind of fractured vision. But putting it this way isn't quite right because the term "fractured" implies the failed possibility of a whole. It is a discursive maneuver that firmly belongs to the downside of the god-trick, to assumptions about the right and proper character of centered knowing and being, to the modern project and its desire for wholeness. So let's say instead that it implies a commitment to sets of partialities, partial connections, and with this, viewpoints of the Other.[5] Another component is the suggestion—indeed the urgent need—that we acknowledge and come to terms, somehow or other, with the specificity of our own knowledges, our situations. Or, to put this in the language of the last chapter, it requires that we explore our own more or less precarious coordination (or otherwise) as knowing subjects. Which returns us to the place from where we started. The issue of the "personal" in academic writing.

Vanity

There's a genre of early-modern painting called the *vanitas,* a painterly meditation on the ephemeral character of the things of this world. Art historian Svetlana Alpers (1989, 103–9), talking about the crafted nature of Dutch painting, describes one such painting: a self-portrait by David Bailly. This particular representation, painted late in the artist's life, shows Bailly as a young man holding a painting of himself as an old man, surrounded by a series of objects that represent the achievements of his life: visits to Italy, artists' materials, and all the rest. These objects are, however, contextualized by others that insist upon the momentary and passing character of the artist's life. So there is a skull, there are flowers which are starting to wilt and drop, and there are floating soap bubbles, beautiful but indeed ephemeral. And, to be sure, there is the juxtaposition of the portraits, young man, old man.

The *vanitas* subsists in a space of tension between the changing person and that which is unchanging and eternal. It plays tricks on, or within, that tension. In depicting it, the painting also *performs* it in its own specific way, specific, that is, to the artistic conventions of the seventeenth century. But if its modality is specific, the divide is not. In one version or another it criss-crosses Western representational forms, this division between, on the one hand, whatever is "personal" and, on the other, that which does not change. For we live with and perform it now, finding it for instance in many forms in our science and social-science writing. It is built into anthropological ethnography, which divides the scholarly monograph from the field notes or the poetry of the anthropologist.[6] It is embedded in the conventions of scientific writing, where literary forms expunge the contingencies and construct a truth that emerges from somewhere outside the specific locations of its production.[7] And it is performed outside the academy, in the division between the artist's personal struggles or circumstances and the eternal character of his insights.[8]

How many times have you heard in conversation between social scientists that to write about oneself is self-indulgent? For, or so they say, it isn't the person who goes and looks, the ethnographer, who is interesting. She may have quirks, and no doubt the fieldwork was messy and embodied, full of problems, hurts, illnesses, and failed love affairs. We all know this, but it isn't important. Rather, it is what is *seen*, has been observed, that is important. What she reports on "out there," to do with Japanese physics, the organization of a biochemical laboratory, or the performance of witchcraft, this is what matters. If we choose to write about ourselves and write "self-reflexive" ethnography, then at best we are getting in the way of what we should be reporting about, introducing noise. And at worst we are engaging in the self-indulgent practice called "vanity ethnography."[9]

The social construction of vanity: that is the distribution being performed in such talk, which turns those who practice self-reflexivity into sites of self-indulgence and presses them to the margins. It is also a method, a performance, no doubt only one of many, that tends to return us to the modern singularity of the god-trick. Indeed, it tends to *enact* the god-trick and, as a part of this, constructs "the problem of the personal" in academic writing, performing the absence of the body from the representation of truth. One might also note, recalling the semiotics of the last chapter, that it rebuilds an ontology of subject versus object, an object out there, prior, something that we may, if we are lucky, come to know.

Here is my sense. Sometimes, perhaps even often, this sneering at the personal is right because the complaints catch something that often doesn't quite work when the personal is introduced into academic writing, when the text starts toward "self-revelation." But the complaint works, or so I suggest, because some versions of self-reflexivity precisely construct themselves as "self-revelations." That is, they play on and further perform the divide between the personal and whatever it is that counts epistemologically, the reports about whatever is said to be "out there."

This poses a problem, or better a task, for those of us who imagine, following Haraway's suggestion, that objectivity (if such there be) is situated and embodied. It poses us the problem of trying to find practices of knowledge-relevant embodiment that don't perform themselves as "self-revelations."

1985: RAF Cosford

I was looking for a subject for study, a case study in the social analysis of technology. I'd done some work in an actor-network tradition on a fifteenth-century technology (Portuguese shipbuilding and navigation) but the sources were poor, and even worse for a nonhistorian, the details of design had been irretrievably lost for the Portuguese vessels in the fourteenth century, no doubt when the craft traditions of the Iberian shipyards in which they were built died out.[10] So I wanted to study a more or less contemporary project and tease out the network of relations, the character of heterogeneous engineering and the malleability of the social. I wanted to explore an approach that insisted the human was no different in kind from the nonhuman. Or, more exactly, that if this were true then it was an effect rather than something given in the order of things. So I was looking for an object of study, but I didn't know what.

One day I took my then five-year-old son for a day out to an aerospace museum called RAF (Royal Air Force) Cosford, which isn't far from where I live in Shropshire. The two of us walked round inside the hangars, looking at the aircraft. Some were civil airliners. Most, however, seemed to be military, ranging from First World War biplanes, through Battle of Britain Spitfires, to examples of some of the more elderly types still in service. The child was pleased with what he saw, and wanted to know how fast each aircraft flew and how high, hoping or guessing all the time that the next one would fly faster, farther, than the one that came before. "Hey look!" he would say, pointing to each new aircraft as it came into view and running off to see it better. I followed more slowly, with a mild resentment at the very fact of being in such a place with its implicit glorification of the military. But I was conscious also of the way in which this resentment butted up against some kind of inarticulate bodily interest in the machines themselves.

Suddenly I saw a familiar shape, the TSR2. I remembered the aircraft well from twenty years earlier. I remembered it because it was controversial for a whole lot of reasons, including its cancellation. So I looked at this aircraft carcass and I thought, "Good God, have they got one of *those* here? Crikey, I didn't know that any of them had survived." And, in the same instant I thought, "That's what I'll study! That's what I'll look at! The TSR2 project."

The Civilizing Process
Donna Haraway has already done it in one way. In talking of situated knowledges, and locating objectivity precisely in the specificities of embodiment, she has offered a particular account of the truth regimes that perform the disembodiment that is normatively required in the modern project. These knowledges belong to what she calls "unmarked subjects," unmarked subjects that turn out to be predominantly wealthy, white, and male. But there are other ways of doing it, other deconstructive stories to tell, not to contradict those of Haraway, but rather to thicken the textures of alternative modes of storytelling, alternative understandings of the specificities of embodiment.

So here is another story from social theory or social history.

When I think of the construction of "the personal," I think immediately of Norbert Elias's work on "the civilizing process" (1978, 1983). It may be that the term itself is not well chosen and that he overgeneralizes, for certainly at times his history seems, well, somewhat mythic. It may also be that he underestimates the horrors produced by "civilization."[12] But does this matter? The answer right now is no, it doesn't. This is because Elias's position is generally deconstructive. It anatomizes the personal and that is what is important here.

Elias says that as the centuries unfold in Western Europe from the late Middle Ages onward, the barrier between the inside and the outside, between the "personal" and the "public" grows. Table manners, bodily functions, the expression of emotions: in the Middle Ages none of these were particularly restrained by comparison with what was to come later. Gradually, he argues, the body and its emotions were concealed behind a wall of politeness, civility, restraint, and repression, and that which was previously visible became private, invisible, inappropriate, unwise. Turned into a matter of no public interest the body became, as we say, "personal."

Elias has a story about how this came about: he says that in times of uncertainty and famine people make the most of the moment. There isn't much point, as I say to my students, in studying for a degree if you know you're going to die of starvation before the end of the year. But this situation prevailed in most, perhaps all, contexts during much of the European Middle Ages. On the other hand, if life is a little more stable—if you think you know where the next meal and the meal after that are coming from—then thinking strategically starts to make sense: thinking and acting long term and not on the spur of the moment. Which means that there is need for repression, calculation, and concealment of one kind or another. This is the point at which dividing the public from the private starts to make sense, for strategy—or so it can be argued—is impossible without concealment.

Elias moves almost imperceptibly between the "social" and the "personal," for in his way of thinking the distinction doesn't make much sense. If people are more calculative because the world is not quite so unpredictable, then this has consequences. It tends to increase the level of predictability yet again, which sets up a virtuous cycle. Stable social life and long-term calculation tend to reinforce one another. This virtuous cycle expresses itself through a range of materialities: the body and its concealments; individual interaction; but also the organization of economic life (a surplus is more likely if social life is calm, but the existence of surplus itself renders social life more predictable); and the organization of the state (which secures a monopoly over the unpredictability performed by violence and therefore, perhaps, secures peace). This is Elias's argument: European history from the Middle Ages onward may be understood, overall, as a virtuous cycle.

Gordon Fyfe suggests that Norbert Elias combines Sigmund Freud with Max Weber, repression with rationalization. This sounds right. What Elias tells us is, to be sure, only a story. Other stories might be and indeed are told. But if our concern is with the archaeology of the personal then it is an interesting story because it tells how what we call the "personal" might have been brought into being and tells, or at any rate implies, that it could be otherwise. And, most of all, Elias's story is interesting because it insists that if we want to understand social life, then we need to attend both to the personal and to the social. Or no: that gives too much away. It suggests that the distinction between the personal and the social is analytically irrelevant.

So what I have told about the events at RAF Cosford in 1985 is a personal story because it is located and makes no particular attempt to perform itself outside time and space, in the eternity depicted by David Bailly in his *vanitas*. I have, I hope, neither pulled any particular god-trick nor tried to perform myself as an unmarked subject. But that story also makes me uneasy. No, better, *therefore* it also makes me uneasy. I think this is because I (and no doubt in some measure many readers) perform the distinction between truth and person, between outside and inside. For most Euro-Americans are the children of Elias's "civilizing process," which means that if we write about "ourselves" then we are sailing close to that divide, the divide that is breached by vanity ethnography or plain, downright self-indulgence. Which is, I think, one of the objections to "reflexive sociology"—one of the reasons that it has not achieved the attention that it no doubt deserves.

Perhaps, then, you are tempted to say: "That's enough of John Law. Enough of the ethnographer. Let's get to the facts! What about the aircraft?" Well, surely this is the issue that we need to confront and deconstruct.

Discourse/Subjectivity

Which story will I tell about Michel Foucault?

Discourses, semiotics, and arrangements of materials of all kinds. Note that: *arrangements of materials of all kinds.*

Talk, forms of storytelling, classical and modern (Foucault 1970, 1972). Systems of knowledge that are embodied in collections, cabinets of curiosities, museums, state records, statistics, doctors' surgeries (Foucault 1976). Buildings, including the shapes of prisons, real and imaginary (Foucault 1979), arcades for visual display, for the gaze,[13] and the Boulevards of Haussman, which cut their clean and ordered way through the pullulating quarters of old Paris. Not to mention the new towns of Morocco.[14]

And then bodies. Yes, for notwithstanding the suggestion that he ignores the specificities of embodiment, the logics of the body itself, there is nevertheless little doubt that Foucault is particularly interested in bodies, bodies and souls. He is interested in how to separate them, how to keep them together, how they are overseen, how they are marked, how they are broken down into little components and then reassembled, pressed into disciplinary forms. How bodies are made in the process of loading a musket in twenty easy steps, walking in formation steady under fire, without the need for further discipline or further orders. And then how pleasures, sexualities, are constructed, pleasures that will normalize themselves and thereby perform disciplinary effects because such are the ways in which the soul, the body, the possibilities of pleasure, have been constructed.[15]

Bodies and souls, and then the other materials: talk, buildings, texts, statistics, maps, plans. Techniques for constituting materials and relating them. For Foucault is a semiotician, and in his archaeology he attempts to decode the logics of relations, the spaces made available by those logics, the spaces, or at any rate the hints of the spaces, denied and made Other by such logics, discourses, or epistemes.

Of course, yes "of course," the distinction between the personal and the rest is of no analytical significance. For the person is, according to Foucault, a subject position constituted in the ruthless logic of a discourse, for instance a disciplinary discourse, while whatever is outside the person is, well, another set of positions that stands in relation to and performs that person, that subject position: for instance knowledge, what is known, or better "knowledges." So the distinctions "public/private" or "knowledge/personal" are *made*, constituted in the enabling logics of discourse that run through, permeate, and perform the materials of the social. They go everywhere, into our bodies, our practices, our texts, our knowledges, our town plans, our buildings, and all the rest.[16]

Foucault's method is quite different from Elias's, but for certain purposes the result is similar. If the truth has nothing to do with the personal and the ephemeral, then that is an effect. It is an effect that fails to notice that the divide is being made continuously through time and through different materials—because the continuities, the logics, the discourses, run through the materials, human and non-human alike.

Interpellation
So semiotics is the study of relations, including the relational formation of the distribution between the knowing subject and the object that is known. Or, if you prefer the language, between the constitution

of the personal, and the knowledges that we have of the world. Michel Foucault was there, a semiotician, but so too was Louis Althusser. And now I want to borrow a term from Althusser: the term *interpellation*.[17] Althusser tells a story about ideological state apparatuses.[18] Talking of ideology, he says that there are moments of recognition, moments when we recognize ourselves because we have been addressed, called out to, in a particular way. At those moments we become, as he puts it, subjects because we are subjected to an authority, a Subject with a capital *S*. We are located, in relation to that Subject, as biddable small *s* subjects precisely because we recognize ourselves and (this is crucial) *because we have no choice.* We are turned into biddable subjects because it becomes instantly obvious to us that we *are* that way and that we *know* that way.

Althusser links this relational semiotics with ideology and its operation through ideological state apparatuses. And, though I want to talk about interpellation, I will have to abandon much of Althusser's project. For instance, his play between Subject and subject: the idea that we are turned into little knowing subjects because we are interpellated by, and mirror, a great Subject. This rests upon the idea that in the last instance there is a kind of ideological coherence, some kind of God eye. Well maybe, but if we take seriously the notion of an established disorder, maybe not. At any rate, this isn't something I want to build into my version of interpellation. I also want to avoid the idea that there are real relations of production that can be distinguished from ideology, that there are, indeed, firm foundations. This, another version of the God eye, makes me uneasy too because, to say it briefly, it's another division that separates appearances too much from reality, the performance of storytelling from what it tells about. To say this is not to say that we will necessarily avoid resuscitating something of this kind that does work. But even so, in a nonfoundational world, Althusser's particular version of the distinction between truth and ideology too will have to go.

But I still want to talk of interpellation because it involves two commitments. First, it is committed to embodiment. Perhaps Althusser doesn't go into this as deeply as Foucault, but even so, it is crucial. For (or so it seems) his sense of interpellation draws from and performs itself through the body of the ex-Catholic. And the ex-Catholic (is there such a thing as an *ex*-Catholic?) knows that he came to believe because he first kneeled and prayed, because he participated in ritual. He also knows that all this happened long before he had any sense of faith, explicit or otherwise. Thus embodiment preceded subjectivity, subjugation to the Subject. We are, says Althusser, interpellated as knowing subjects precisely because we *are* embodiments, embodiments of relations and gestures.

Second, I want to hang on to Althusser's insistence on obviousness. For me this is the fulcrum of interpellation: *the subject instantly recognizes itself when it is addressed.* Note that: the subject *instantly* recognizes itself and is constituted as a knowing subject when it is spoken to. Indeed (and he is equally insistent on this) the constitution of the subject *precedes* the words spoken, the fact of being addressed. So interpellation has nothing to do with "deciding." All the apparatus of "rational decision making" (assuming we believe in the existence of such a beast in the first place) is bypassed. Instead there is instant recognition and location.

Perhaps Althusser was thinking of words, words and bodies. Perhaps he was imagining the words of the priest, the schoolteacher, the politician, or the bourgeois political economist, the effect of all these words on the body of the subject. But there is no reason to restrict interpellation to words. Indeed, the emphasis on embodiment suggests that words are at best only the beginning.

Interpellated

Now I want to narrate a story that joins the personal to that which is not ephemeral, the subject of study with the object of study. This is because I don't think that the personal is "personal" when it is put in this way. But we've needed Alpers, Elias, Foucault, and finally Althusser to reach the point where it is possible to theorize the personal in a way that resists its designation in those terms.

1985. RAF Cosford.

It's like this, isn't it? This was a *moment of interpellation.* Knowing subject, known object, the two were recognized together in a single instant. I would study, study the TSR2; the TSR2 would be studied, studied by me. The effect was, as it were, an instant recognition or performance of a set of subject/Subject or subject/object relations coming from—well, coming from somewhere, but deeply buried in its obviousness, somewhere before.

Obviousness? Let's just remind ourselves of some of its dangers. If we are interpellated, then we are being made or remade as a particular subject position, made to constitute our objects in particular ways. In particular we are being made to constitute our objects in ways that are obvious, recognized and made before we come to see them and think about them. There is another study here. We might think of it as the erotics of interpellation: Why or how it is that we are spoken to and perform the obviousnesses of our objects of study?[19] Technoscience studies, military technologies with all their genderings, biomedical this and that, consumer goods, in all of these we are making obviousnesses of one kind or another. I say that we are *making* obviousnesses because our narratives are performative. But if this is the case, then the question becomes: Interpellated as we are, what on earth is it that we are performing in our embodiments?

Large Blocks

Michel Foucault and Louis Althusser have something in common other than their radical politics and their semiotic interest in relations. Both tend to imagine that the logics of discourse, of semiotics, come in very large chunks. Althusser, talking of the multiplicity of and differences between ideological state apparatuses (ISAs), writes: "If the ISAs 'function' massively and predominantly by ideology, what unifies their diversity is precisely this functioning, insofar as the ideology by which they function is always in fact unified, despite its diversity and its contradictions, *beneath the ruling ideology,* which is the ideology of the 'ruling class'" (Althusser 1971, 139). And Foucault? "In the last years of the eighteenth century, European culture outlined a structure that has not yet been unravelled; we are only just beginning to disentangle a few of the threads, which are still so unknown to us that we immediately assume them to be either marvellously new or absolutely archaic, whereas for two hundred years (not less, yet not much more) they have constituted the dark but firm web of our experience" (Foucault 1976, 199).

"Is always in fact unified." "The dark but firm web of our experience." So the metaphors and the theories are different, but they have this much in common. Despite the cracks and the strains, most of the space—the space made by ideology for Althusser and the episteme for Foucault[20]—most of the space we have for knowing and being, living and building is structured by a single set of ontological strategies or distributions. These distributions (this is the crucial point) displace others to the margins, to the places that Foucault, with his exquisite sense of spatiality, sometimes calls the heterotopic.[21] They are displaced to a few places of resistance. Althusser again: "I ask the pardon of those teachers who, in dreadful conditions, attempt to turn the few weapons they can find in the history and learning they 'teach' against the ideology, the system and the practices in which they are trapped" (Althusser 1971, 148).

This is okay intellectually if not politically. Perhaps it is really like that. Perhaps, to use Donna Haraway's very different words, we live in a world dominated by narrative metaphors that perform "militarism, capitalism, colonialism and male supremacy." Perhaps we live in a world in which the narratives that perform one another tend to support one another despite some discordances and thereby produce such obnoxious singularities. Perhaps we live in a world in which the "personal" subject positions into which we are interpellated are, indeed, more or less consistent, performed as reasonably coherent and well coordinated locations within a single great episteme. Perhaps we live in a world that presses toward singularity rather than multiplicity, a world that is modern rather than postmodern in character.[22] If it is like that, or largely like that, then an inquiry into our interpellations as knowing subjects seems necessary, indeed vital. That is, if (like Althusser's teachers) we can manage "to turn the few weapons [we] can find . . . against . . . the practices in which [we] are trapped" and so make different stories and narratives, narratives that explore the ways in which our constitution as subjects has generated objects and knowledge relations that perform their obviousness under whatever sign "science" and "reality" now sail.

Perhaps. But I am just a little more optimistic than that because I take it that the established disorders are multiple as well as singular. Notwithstanding the coordinations of the many strategies for coordination, the strains toward singularity are counterbalanced by the heterogeneity of multiplicity, and the interferences and overlaps do not necessarily ensure that the fields of ontological force are aligned. I am more optimistic because I take it that the conditions of possibility do not necessarily come in large blocks.[23]

1985–1989

Between 1985 and 1989 I studied the TSR2. I sought to develop actor-network theory by adding the TSR2 to the long list of technoscience case studies: bicycles, electric lighting, electric vehicles, electricity power systems, imperialist ships, turbojets, Pasteurizations, aqua-cultures, domestic heating, technology transfers. I was specially concerned with the social and the technical. The social, or so I sought to show, was as malleable as the technical. Its shape was not given in the order of things but rather emerged in the course of "heterogeneous engineering." To show this I told stories to do with the trajectory of the TSR2 project, its rise, its development, and its fall. I also told stories of particular specificities in this trajectory, particular incidents. I was quite pleased with some of these project stories. In the space created for this new sociology of technology they seemed to work reasonably well.[24] But things started to go wrong. This showed itself in a number of ways. In particular, however, I started to become uneasy about my relationship with this aircraft, and especially, with its high-status spokespersons. These were people who were often impressive, thoughtful, successful, and attractive. But as I talked to them, I gradually found that these conversations seemed to be laying a charge on me: I was being constituted as the person who would document this project "definitively." It was, or so some of them said quite explicitly, the right time to document the project. Far removed from those sad days in 1965 when the aircraft was canceled, people could now look back calmly and reflect upon what had happened. By now mostly in retirement, enough of them were still alive to make the study possible.

And then this. I wanted to write an actor-network story, a narrative exploring the malleability of the technical and the social. But my interviewees knew nothing of actor-network theory. Instead, they treated any empirical description as an admittedly crucial prelude to the more important task of making balanced judgments—and two such judgments in particular. First, whether it was indeed right to have canceled the TSR2. And second, and more generally, what might be learned from this sorry story that could be applied to other military aircraft projects.

I had written most of a book on the TSR2 project. But toward the end of 1989 I put that manuscript in a box, put the box on a shelf, and went to work on a quite different project. And that manuscript is still in its box.

Five Narrative Forms, Five Interpellations

In order to make sense of all this I now want to tell a rather formal story. I want to imagine that we are concerned with five discourses, five separate discursive forms, five distributions, five modes of interpellation. Putting it in this formal way means that the differences that I make are too discrete, too clear, too abrupt. In practice, matters would always be more subtle, less clear. But I do so in order to make a simplicity that will help in the process of exploring a logic of interference.

The discursive forms.

Let me call this first form history, *plain history*. It is a story, a form of storytelling, that starts at the beginning and moves to the end. *A* follows *B* follows *C*. It is a story that charts the inception, the conception, the development of a project, its growth, and its decline. Perhaps it charts its cancellation. At any rate, it is a story that tells of the trajectory of a technoscience project. This small example of the genre comes from a book by a well-known military aviation journalist and commentator, Derek Wood.

> By mid 1957 the RAF had formulated its basic requirement under the title of General Operational Requirement No. 339 and it was passed to the Ministry of Supply for action. The Controller Aircraft, Sir Claude Pelly, sent out a letter to industry on GOR.339 on September 9th. (Wood 1975, 153)

There is obviously no such thing as "plain history." All history, "plain" or otherwise, is a narration and a performance. It makes, distributes, and links things together, bringing them into being and asserting their significance (or otherwise) by chaining them into possibly chronological sequences. I'll talk more about the properties of such "arborescent" arrangements in chapter 8. For the moment just let me observe that it performs them too, constituting some kind of truth regime and, no doubt, effecting some consequences. This little excerpt is just that, an excerpt. It makes one or two links in a narrative that would be much longer if we were to spend more time on it. But in the present context this does not matter, for my point is that in many stories of social science—including those that have to do with technology—there is this sense of something like a lowest common denominator: the making of a series of linked dates and events. This, then this, then this. The effect is the production of something, a set of specificities, specific object positions, which will often subsequently come to act as the "raw material" for other forms of storytelling, other discourses. And at the same time, a set of more or less centered subject positions or a reader/author is produced that makes, that appreciates, that is interpellated, by "the facts as they are"—the facts, for instance, about a project, an aircraft project.[25]

I'll call the second discursive mode *policy narrative.* This mode tells a story that has something to do with the first narrative, that of plain history, but in policy narrative the specificities are distributed into chains that are energized and achieve value by being given some kind of pragmatic policy relevance. That is, the specificities that might have been built in "plain history" are now awarded the potential for judgment or for contributing to a judgment. This, then, is a normative form of narrative, one that chains itself together by distributing praise, blame, and responsibility. It is energized with polarities, with pluses and minuses. And it creates a series of engaged, normative, and more or less centered subject positions. Here is a small example of the genre.

> Sixth, and probably the greatest single cause of increased costs, was the repeated delay in getting official decisions, and the permanent uncertainty which so grievously effected the rhythm of production and from time to time the morale of design and production teams. (Hastings 1966, 60)

So this is an instance of policy narrative, a story about TSR2, written by a historically minded policy practitioner, Conservative Member of Parliament (MP) Stephen Hastings. And the subject made in this form of writing is that of the judge, a location or a role in which the pros and cons, the lessons, may be weighed up.

Narrative style number three is somewhat like number two. It is about judgment, right and wrong. In particular, it is a way of talking and living that is *ethical* in character. Perhaps, then, the distinction is like that made by Max Weber between instrumental and value-rational action. (See Weber 1978, 24.) Here, at any rate, is Labour MP Tam Dalyell in a parliamentary debate that took place immediately after the cancellation of the TSR2, a censure debate:

> It is a sombre fact that as the twentieth century rolls on more and more science-based developments have evolved from armaments. The right hon. Gentleman is, tragically, historically accurate when he says that armaments have been the life-blood of industry, but the fact that he is historically right casts a pretty damning reflection on contemporary capitalism. . . . One of the reasons why I became a Socialist was my belief that in a Socialist set-up at least one had a chance of creating conditions in which technical progress could be freed from the armaments race. (*Hansard* 1964–65)

Wertrationalität rather than *Zweckrationalität*. Here the links that provide for the performance of the narrative are normative—but also ethical. The subject positions are interpellated and linked together by means of a particular sense of right and wrong. Interpellation—and relative subject coherence—are achieved by moral means.

Narrative number four is different again. This I will call *esoteric narrative* (though perhaps all narrative styles are more or less esoteric—or at least more or less specific). This is because I want to find a way of talking about forms of storytelling that are both local and self-professedly analytical—as, for instance, is the case for most versions of academic storytelling. There are, to be sure, innumerable variants here, just as there are innumerable variants of policy storytelling and, for that matter, of historiography, plain history. But sometimes, perhaps often, academia is a place that makes stories, but stories where the links, and so the reader positions, are indeed esoteric in quality: interpellative for a specialist reader, and drawing on but not very directly feeding back into other forms of narration. And it is this sense of removal, of strictly local relevance, that I want to catch by talking in this way. So the story I am telling right now (though it performs arrangements that will in some measure interfere with other genres) is no doubt a version of esoteric narrative. But so too is this:

> At this point, then, the project came properly into being. The managers had been granted an area of relative autonomy by ac-

tors in the global network; they had been granted what we will call a *negotiation space* in order to build a *local network.* (Law and Callon 1988, 289)

This is Michel Callon and John Law working with actor-network theory, making arguments that operate in a more or less local space. And again, as is obvious, there is a relationship between this narrative and that of plain history. The one is, in some sense, parasitic on the other.[26]

And number five? This is rather different: let's call it *aesthetic.* It is a form of narrative that has to do with distributing and performing pleasure, with what is beautiful. It makes, in short, a particular form of (no doubt gendered) aesthetic subject.[27] This is an issue that I will look at more closely in chapter 6. For now, however, consider the following:

> The cockpit felt almost detached from reality. There was no vibration; the only noise a subdued hum from the big turbo jets, the seat comfort as luxurious as an airliner's and the air conditioning warm, fresh and comfortable. The instrument panels showed steady readings of flight conditions, engine and systems performance with no malfunctions; the radio for once silent.
>
> Outside an unbroken cloud sheet stretching below to the pale northern horizon, varied in colour only here and there by long streaks of shadow laid by the low winter sun from behind strato cumulus domes.[28]

The excerpt is from a book by the test pilot Roland Beamont, and much could be said of it. For instance, like David Bailly's *vanitas* it artfully bridges the distinction between public and private, but at the same time it also performs it. The pilot, or so we learn, is a cool technician ("the instrumental panels showed steady readings," and he recognizes "strato cumulus domes"). But there is also poetry here, poetry, pleasure, and no doubt sadness too, the sadness of a loss that will come, as the TSR2 is grounded and destroyed. This, then, is a performance of a particular form of aesthetic narrative—another interpellative logic, the production of another subject position.

Interferences

So we are dealing in multiplicity. Multiple forms for object positions. Multiple styles of subject positions, interpellations. Donna Haraway (1994, 62) observes: "Optical metaphors are unavoidable in figuring technoscience." And a few sentences later notes that "my favourite optical metaphor is diffraction—the noninnocent, complexly erotic practice of making a difference in the world, rather than displacing the same elsewhere" (Haraway 1994, 63).

Here I have made five forms of narrative, suggested five styles for making the interpellations of subject positions and object positions, five modes of distributing. These are the story forms that I have identified: plain history, policy, ethics, the esoteric, and the aesthetic. I have, to be sure, made them unrealistically discrete. I wouldn't propose for one moment that they are fixed, unbridgeable, or primitive. And even if one sticks with the labels, clearly they don't represent invariant and unchanging modes of interpellation. They are products of particular material-semiotic circumstances. But in the present context I have set the narratives up in this way, as discrete and separate, because I am interested in the ways in which they interfere with one another. In particular, this makes it easier to see that though they may individually make more or less coherent subject positions, something different starts to happen when several of them are juxtaposed. For when they start to interact with one another to generate complex patterns of interference, they also start to make subjects, readers, and authors that may be places of illumination when the wave patterns are coherent, but which may be places of darkness when they are not— when the wave patterns cancel one another out.

It is, of course, possible to make the argument in many ways. But here I want to link it to the formation of subjectivity. To the formation or the performance of what I hope we are no longer so disposed to think of as "the personal."

So the story runs thus. In 1989 these patterns of narrative coalesced in a particular way to interfere with one another and make a place of darkness. They overlapped to produce a series of mutually destructive interpellations, conflicting subject positions, a place where there was no possibility of writing, reading, or knowing.

I was writing history. This was the first interpellation. But it wasn't "just history," not in the way I imagined it. For plain history was never the strongest form of interpellation for an author made as a sociologist or a student of technoscience rather than a historian. So it was a prelude to what? I thought I was writing history as a prelude to a particular academic narrative, a story about human and nonhuman agents. So that was the second interpellation, the recognition or construction of a subject by an esoteric discourse: in this case the narrative of actor-network theory. But my high-status interviewees imagined that I was writing history, making my inquiries as a prelude to something quite different: a judgment of policy, a judgment about the worthiness of the project. So they thought I was going to write about whether or not it was well managed, what had gone wrong, why it had been canceled and, perhaps most important, whether or not it should have been canceled. This was a third interpellation and it was one that was powerful. It was an effect of what is sometimes called "studying up" and felt like a form of colonization.

So there were three simultaneous interpellations—but two of them were interfering with one another. The writer constituted as esoteric specialist and the writer as policymaker; in this context the two did not fit.[29] So that was the first interference pattern. A place of darkness. But it was not the only one. There was also a resistance deriving from another story form, a long-ago story form, a story to do with the waste of military spending. This was a form of being that reflected and embodied a horror of nuclear weapons. Perhaps it was an ethical discourse that had been almost buried, long ago cleansed from other narrative forms, from those to do with either the esoteric or matters of policy. So what form did the interference take? I found I *could not* perform policy. I could not, that is, perform the kind of policy that judged the TSR2 in terms of its (how to say it?) efficacy with respect to military strategy or procurement policy. This was the point at which I stopped. I could not make that final move. Interpellation as a policy narrator interfered, though in different ways, with *both* esoteric and ethical subject positions.

Interpellation, diffraction, and interference, a moment of darkness, destructive—or at least immobilizing—subject multiplicity.[30]

Can we imagine this as "inauthenticity"? I don't know, for certainly in a world of multiplicity there is no "core," no last instance, no soul, no ultimately centered subject. Which means that authenticity, if that is what it is, rather has to do with the relations *between* narratives, of the relations between different forms of obviousness, between the obviousnesses performed in different but related subject positions.

And what of the question of aesthetics? This was another form of interference, I think, for there is a pleasure in aircraft.

No. Let me be careful. Some of us, some of the time, some subject positions, are constituted to find various kinds of pleasure in relation to machines, flying machines, and even killing machines. That is a whole other set of narratives, ways of talking and being that are pressed, or so I'd want to suggest, to the margins, in the narrative forms that I've mentioned above. My conclusion is that some version or other of machinic pleasure was what interpellated me one day, in RAF Cosford, in 1985, and then hid itself again. Which is some kind of answer as to why I "chose" to study TSR2 when I might have chosen to study so many other technologies.[31]

To End

I started this chapter with two beginnings. One had to do with multiplicity, indirection, and the coherence of subject positions. The other concerned the reflexive problem of the personal. Now it is possible to say that they overlap and interfere with one another.

Some observations.

I want to assert, against the purifying tropes of modernism, that *whatever is personal is also social*. Always. Whatever we conceal, we think is shameful, inappropriate, self-indulgent, uninteresting, whatever we conceal is *also* social. Elias tells us this.

It may, of course, *also* be shameful, inappropriate, self-indulgent, or plain, downright uninteresting at the same time as being as social. All of these are real possibilities which we watch being performed every day. They are, indeed, performed in one way by Elias and Foucault in their writing—if only because they choose never to talk about their own "repressions," their own subjectivities. But if we assume, as I have in this chapter, that narratives perform subject positions and object positions, then there is, at least in principle, the possibility that subject positions, those positions that constitute us as knowing subjects, are relevant if we want to understand the performativity of narratives, to understand how distributions are being made, if we want to understand what is being said and what is not.

So that is a first possibility. There is continuity between subjects and objects, and we are lodged in, made by multiple and overlapping distributions, which shuffle that which is made "personal" and that which is rendered "eternal" into two heaps. And that process of shuffling is worthy of deconstructing in a contemporary version of the *vanitas* because it performs obviousnesses. And *in particular* because it performs obviousnesses.

This is where, having journeyed almost all the way with the semiotics of Michel Foucault and Louis Althusser, I finally part company from them. This is a methodological point. For I would argue that the body is a particularly sensitive instrument in part precisely because the semiotics of subject-object relations *don't* come in big blocks like ideologies, discourses, or epistemes. Let me be more cautious. They *may* come in big blocks in certain respects. Perhaps the conditions of possibility are, indeed, in some ways uniform, singular. But they aren't that way all the time. For smaller blocks, narratives, semiotic logics, distributions—these are multiples that are also capable of interfering with and eroding one another. At any rate they are capable of doing so under certain circumstances, in particular places or institutions. They can produce multiplicities that do not effortlessly coalesce to make singularities. In, for instance, the practices within a building, the intertextualities that pass through the body, the heterotopic space within that makes us, interpellates us and our materials in multiple ways.[32]

Lighten our darkness. Deliver us this day from the obviousness of our simplicities.

This is why I am more optimistic than Louis Althusser. There is mileage to be gained by attending to interferences that make multiplicities. And it is also why the body is so important. For it is a detector, a finely tuned detector of narrative diffraction patterns. It is an exquisite and finely honed instrument that both performs and detects patterns of interference, those places where the peaks come together and there is extra light. And those, such as the place I found myself in the summer of 1989, where there is dark, where there is something wrong, where the energies cancel one another out. Where multiplicity is not reduced.

This suggests that there is a place for the body, not only as the flesh and narrative blood that walks in what we used to call "the field" bringing back reports, reports of how it is "out there," but also that there is room for the body, for the personal, in the narratives that are later performed, that perform themselves through us as we tell of narrative diffractions and interferences. For the personal, when we come to sense it in this way, is no longer "personal." It is no longer necessarily personal, however it may be constructed by the modernist-inclined heirs to the civilizing process. It may be understood and performed rather as a location, one particular location, of narrative overlap. A place of multiplicity, of patterns, of patterns of narrative interference. And of irreduction.

Whether we tell stories about ourselves as we perform our situated knowledges will depend on what we are trying to achieve and on the context in which we are seeking to achieve it. The performance of reflexivity and diffraction does not necessarily demand the immediate visibility of a narrator. But the issues are situated, specific, rhetorical, and political in character rather than great issues of principle. For it is itself wrong, a confusion, a self-indulgence, to forget that the body is a site, an important site, where subjectivities and interpellations produce effects that are strange and beautiful—indeed sometimes terrible. And these are effects that might make a difference if were able to attend to their intertextualities.

For instance, there are moments—I lived through one that I have already described—when the possibility of performing coordination between narratives is lost and it is no longer possible to link subject positions together in this way or that, to make a single story; when it is no longer possible to create, perform, and be performed by an object that is turned into a singularity; when it is no longer possible to work, as it were, perspectivally. In such moments, the interferences and overlaps perform themselves into "a" subject that is broken, fragmented, and decentered; a subject that is therefore interpellated by—and interpellates—a multiplicity of different objects and thereby suddenly apprehends that the failure to center is not simply a failure but also a way of becoming sensitive to the multiplicities of the world. At that moment, failure to center is also a way of learning that objects are made, and that there are many of them. It is a way of learning that objects are decentered—a set of different object positions—and a way of attending to the indirections of interference. It is also a way of apprehending that knowing is as much about making, about ontology, about what there is, as it ever was about epistemology.

> **A White Bird**
>
> Years later, in March 1996, I looked at a videotape of the first flight of
> the TSR2, a version of the publicity film issued by the British Aircraft
> Corporation in 1964. The result was unexpected because it was thrill-
> ing. It was thrilling to see it start down the runway. And then, with a
> gap (for the film was not technically outstanding) watch this aircraft
> take to the air like a great white bird. Perhaps it was the music, for
> they played the theme from the film *Chariots of Fire*. Perhaps.

A final question. Could I have done all this without introducing the
personal?

The answer is no. Perhaps I could have made arguments like these
and *told* it otherwise in some version or other of the god-trick. But
this is not how the method of bodily interference produced its effects.
So I'll finish with another question. If we are constituted as knowing subjects, interpellated, in ways that we do not tell, then what are
we doing? What are we telling? What are we making of our objects of
study? Or, perhaps better, what are they making of us?

The question is real, isn't it? At any rate it's real from where I stand.
For finally, in a study of the TSR2, it turns itself into something specific that is also not specific. If those of us who study military technologies—and those who dream of them, design them, fly them—do
not reflect on the aesthetics of our interpellations then we are not attending to a way of living stories that runs through us. A way of living
stories that is arousing, in some ways dangerously so, that effaces the
ontological in favor of the perspectival, and that makes a difference
and continues to strain toward the singularities of military and technological discourse.[33] This is the power of a reflexive technoscience
studies: it can attend to, and learn from, dangerous arousal.

Multilingualism is not merely the property of several systems each of which would be homogeneous in itself: it is primarily the line of flight or of variation which affects each system by stopping it from being homogeneous.
—Gilles Deleuze and Claire Parnet, *Dialogues*

So there are multiplicities. There are multiple distributions of subjects and multiple distributions of objects. And these distributions overlap. Sometimes the overlaps work to make patterns of light, somewhat singular narratives. Sometimes they consolidate themselves to make coherences, simplicities. And sometimes they do not —and then we find that we are left in the dark places, turned into a fragmented set of subject positions confronted by an equally uncoordinated set of object positions.

No doubt this is uncomfortable. But, if we can work it right, perhaps in those dark moments it is easiest to learn about the making of objects and the making of subjects because in those moments it is easiest to attend to the work of distribution and coordination. And, in particular, those are the moments when it is easiest to avoid being dazzled by problems of epistemological authority and deal, instead, with ontology: with the making of what there is or there could be, with the conditions of possibility. With the performances that otherwise tend to reenact singularity. This, then, is the interest in interferences, that they allow us both to rethink and survey the character of distribution—with how it is that matters are made and arranged in the world.

In this chapter I follow Sharon Traweek and tell more stories while looking for the distributions that they make. I also follow Annemarie Mol by attending to the ways stories describe and make links: connections and disconnections or similarities and differences—that is, by attending to their interferences.[1] I argue that to tell stories is to perform "cultural tasks." It is to distribute, to say what exists or does not. And it is to coordinate, to say what goes with or does not go with, what else. This means that I'm assuming, as I have above, that storytelling is performative: it makes or may make a difference in the multilingual world mentioned by Gilles Deleuze and Claire Parnet. Talk may, as it were, talk itself into being, and the stories told may shift their material form, may perform their logic from texts or voiced words into bodies and architectures, into other forms of flesh, and into stone. So the echoes here are also with that material version of semiotics called actor-network theory and with the understandings in cultural anthropology or cultural studies of the ways in which, for instance, communities may be imagined and told into being as different stories overlap.[2]

English Electric

English Electric: in the 1940s and 1950s this was a proudly indepen-dent company based in and near Preston, a large town north of Man-chester in Lancashire in the UK.

A brief history of English Electric? The company was a success-ful Second World War aircraft manufacturer. It worked by taking the designs of other companies and producing them under license effi-ciently and on time. This was fine for wartime because the United Kingdom needed all the aircraft it could get, and it needed manufac-turers even if they didn't design their own aircraft. But at the end of the war, the directors could see that if the company was to survive as an aircraft manufacturer, it would henceforth need to create its own aircraft from scratch. So in 1945 it created its own design team.

The new team knew that they only had one chance. If they got it wrong, English Electric would have to make do with manufacturing industrial machinery or domestic appliances. So it needed to design an aircraft that would be attractive and would sell. This meant, in par-ticular, that it should be cheap, flexible, reliable, and simple. So, bor-rowing the technology of the defeated Germans, the company built a light bomber and reconnaissance aircraft. Straightforward, subsonic, but immensely versatile, it was code-named the Canberra and turned out to be a world-beater. It sold in thousands, both to the Royal Air Force and overseas, and was manufactured under license in large numbers abroad.

The gamble had paid off. English Electric was successfully estab-lished as a front-rank aircraft manufacturer. But what should follow?

At this point there was a disagreement between the English Electric designers and the Whitehall civil servants who were responsible for British military aircraft procurement policy. The mandarins thought that supersonic technology was risky, that it wouldn't pay off, so they continued to order subsonic aircraft. At English Electric they thought differently, and putting their money behind their ideas, they designed and built a prototype supersonic fighter aircraft, code-named the P1. In some ways this was a tricky machine. It wasn't easy to service, its move into production was beset by delays, and it carried little fuel so its range was very limited. But in other ways it was extremely successful. In particular, it flew brilliantly. In the end Whitehall came around and bought a developed version of the P1, called the Lightning, for the Royal Air Force. And, though it didn't match the extraordinary success of the Canberra, the P1 also went on to sell very well overseas.[3]

Two out of two: the Canberra followed by the P1 Lightning. English Electric had become a very successful aircraft company. But what would follow the P1?

We have reached 1955 now and find that the Royal Air Force was thinking hard about its future aircraft. Here's an excerpt from a confidential government memo:

> The Canberras, with the ability to deliver the tactical atomic bomb and trained to operate at low level, must continue to provide our tactical strike and reconnaissance force for some time to come. It is difficult to say how long they can continue to be regarded as an effective tactical force. However, operated strictly at low level, they might perhaps continue to do so until the enemy can develop an effective low level surface to air guided weapon. At best this might be until 1963. (AIR8/2014 1955)

So there was a gap, a space for a Canberra replacement. It was a space defined by the threat to subsonic, medium-altitude bombers flying over Russia posed by antiaircraft missiles which might shoot them down. And it was a space that gradually took shape between 1955 and 1957 when it was specified in a document called General Operational Requirement (GOR) 339. This is what English Electric was after: the contract to design and build the GOR 339 aircraft, the Canberra replacement.

It's possible to tell a story about the evolution of that design, the steps the English Electric designers went through.[4] By 1957 this design had stabilized in a particular proposal code-named the P.17A. This design was described and justified in a long brochure written in response to the Whitehall requests for designs for a GOR 339 aircraft. Most of the brochure is given over to technical description of one point or another. But it also contains a history or perhaps it would be better to say a genealogy of the P.17A, which was, so to speak, a description of its antecedents.

> The value of the Canberra experience cannot be over-estimated. It is the only modern tactical strike and reconnaissance aircraft in service with the R.A.F. and many other Air Forces. More Canberra aircraft are in service with foreign countries than the Viscount, which holds the record for British civil aircraft. This is due to the flexibility of the Canberra in its operational roles and performance, and is a factor which has been kept in mind throughout the P.17A design development.[5]

In this excerpt from the brochure we're not only being reminded of the history that I have just recounted but also (in a version of the policy genre discussed in chapter 3) of its relevance. For the Canberra, or so the document is going to tell us, is an excellent test bed for all the tactical equipment needed for the new aircraft—the radars, the bombing equipment, and all the rest. The Canberra also has the virtue that it does some of the same jobs that the GOR 339 plane will do: "the Canberra is being used for low level strikes with delivery of tactical atomic stores by L.A.B.S. manoeuvres" (English Electric/Short Bros. 1958, 1.S.3). LABS is an acronym. It stands for Low Altitude Bombing System, which is a term that describes the maneuvers the plane goes through in order to avoid destroying itself as a result of delivering "tactical atomic stores."

So the Canberra was some kind of progenitor. But in many systems of kinship, offspring have two parents and the P.17A was no exception. So we move to more history, or more context.

> Meanwhile the P1.B is the only aircraft under operational development having high supersonic experience and appropriate auto-pilot and instrument systems. Moreover, it is the first aircraft under development as an integrated weapon system with all-weather equipment and a reasonable degree of automaticity.
>
> Perhaps most important of all, it is the only aircraft in the world known to have flown with satisfactory controllability up to a

Mach number of 1 at very low altitudes in very rough air. (English Electric/Short Bros. 1958, 1.S.3)

So the argument was that the P1.B already flew like the OR 339 aircraft, very low and very fast, and it did so well. The promise was there. The experience of the P1.B would be built into the P.17A. More lines of descent. And what this particular passage doesn't mention (though it crops up in the narrative elsewhere) is that many of the techniques used to design the P1.B were also being used for the P.17A. The P1.B stress calculations, for instance, were run on a big computer, the DEUCE, which was purchased for the P1.B project. Now the same computer was being used to design the P.17A, not to mention the high-speed wind tunnels and all the accumulated design office experience.

The brochure adds the following:

It will be seen that the P.17A represents a completely straight-forward application of our design experience, as of 1957, just as the Canberra was a conventional application of aerodynamic and structural design knowledge in 1945. This is for the same reason; to guarantee that the R.A.F. have a practical aircraft in service as near as possible to the desired time scale.[6]

Continuity and Culture

I will shortly take this story one step further, but it is time to pause for a moment and reflect. Let's start by saying that this history as told by the brochure writers at English Electric is perfectly plausible. No doubt as an expression of the coordinating potential of what in the previous chapter I described as "plain history," it might be incorporated into an account offered by any historian of the English Electric company, and I suggest, its general style feels comfortable in the context of technoscience studies. So what is the nature of this plausibility?

The answer lies in the fact that one thing leads discursively to another. Somehow or other, events go together, distributed onto a line, a time line, a line of influence, the teleological means-ends line that is the guiding thread of a project. It is, to be sure, an interested history. It would be easy to tell a debunking story about the concerns of those who wrote the brochure, noting that they sought to make as much as possible of the readily available cultural materials. And it would be equally easy to tell a story that did not debunk but merely noted the operation of social interests as well as the existence and manipulation of a prior set of resources in the form of skills, materials (such as wind tunnels and machine tools), and texts.

I'll discuss interest narratives in a later section of this chapter. But such ironicizing or contextualizing doesn't necessarily reduce the plausibility of the story about English Electric. Note, for instance, that it conforms, at least in broad shape, to the form of much narration in technoscience studies, sociology, or anthropology. It does so, in particular, because it is an example of an origin story.[7] The narratives retell how one (cultural) thing leads to another, influencing it and shaping it, as one passes through time. So it is a narrative in a plausible form, in one or more of the versions of that form ("plain history" and "policy narrative")—or their closely related if more esoteric cousin, the social shaping of technology. It makes a reader who knows how to handle and assess it, who knows the strategic moves. In addition, that reader knows the kinds of issues that might be highlighted if one wanted to set about undermining it: "We need more detail" or "No, the similarities between the P1.B and the P.17A are overstated if we look at this other material." And so on.

So, let's say that this form of narrative is a coordinating strategy, a method for the cultural ordering of what might otherwise be disconnected objects. It takes the form of a plausible historical narrative, a plausible origin story. It makes a culture (we perhaps should remind ourselves again) that ramifies into and is performed through material objects and procedures such as genes, skills, jigs, and power presses, a culture that somehow or other may be said to shape the events that it contains, in this case historically.

So what about this term, *culture*? It would, to be sure, be possible to write a book about this. Indeed a library. Several have been written. I want, however, to approach the term in a particular way by linking it to specific lines of writing in technoscience studies. With this in mind, it is helpful to cite Sharon Traweek, who tells us that "a community is a group of people with a shared past, with ways of recognizing and displaying their differences from other groups, and expectations for a shared future. Their culture is the *ways,* the strategies they recognize and use and invent for making sense" (Traweek 1992, 437–

38). So we have strategies for arranging, for making sense and (to add to her definition) *creating* similarities and differences, including the similarities and differences that constitute community.

But how are similarity and difference made? As I suggested in chapter 2, there are various strategies, methods for distributing or ordering such continuities and ruptures. And here we are dealing with another: that of *chronology or genealogy,* the tracing of descent, the insistence on commonality through the generations. This strategy comprises at least one of the tropes used by those who wrote or (we might add) performed the English Electric brochure; by those who worked in the test facilities and factories of English Electric in the north of England at Warton and Preston; by the material embodiments of English Electric, precisely in the form of those facilities and factories; by the story, by the lineage, that I built for the company at the beginning of this chapter; and by technoscience students as they seek to make sense of the way in which things follow things to produce what we might think of as *shaped continuity.* For this is one of the great distributive tropes, methods, or mechanisms of culture. It is not surprising that we should find it in our materials. It is not surprising that we should use it ourselves in our technoscience studies: the narrative of the world as genealogy and chronology. More time lines. A project doesn't need to be made in this way. No doubt it cannot exclusively be made in this way. But surely this is one of the elementary mechanisms of project making.

Vickers Armstrong

There's a nasty dig in the English Electric brochure that I mentioned only in passing. This has to do with the Viscount aircraft, which hasn't sold as well, or so the brochure claims, as the Canberra. It is a nasty dig because it is a way of making a difference between English Electric and one of its rivals, perhaps its major rival, the manufacturer of the Viscount aircraft, a firm called Vickers Armstrong.

For English Electric was not alone in hoping to win the GOR 339 contract. A host of other companies were jostling for a piece of the action,[8] and one of these was Vickers Armstrong. It was really two firms. One was based at Weybridge in the southern suburbs of London. The Weybridge firm was in the process of digesting another based in Hampshire, in the south of England, called Supermarine. I'll have something more to say about the effects of this merger later, but for the moment just let me say that Supermarine, which had manufactured the Second World War Spitfire fighter aircraft, had a design team that thought very much in terms of integrated systems, while Vickers Armstrong was a major producer of successful aircraft, both commercial (names like the Viking and the already-mentioned Viscount belonged to Vickers) and military (the first of Britain's strategic nuclear V-bombers, the Valiant).[10]

So Vickers Armstrong was a highly plausible contender for the GOR 339 contract:

Meanwhile, Vickers Supermarine had been working on a number of alternative designs, the vulnerability of which had been carefully tested against the ideas of Vickers Guided Weapons Division. Their experience was limited to transonic aircraft. . . . The various designs were submitted to a cost-effect examination against GOR 339, and as a result Vickers tendered first for a small single-engined plane suitable for both the Air Force and the Navy. The design was in the Supermarine Spitfire Tradition. (Hastings 1966, 30)

So here we see another historical story, the construction of a series of similarities, descents, that take us back across time to what is perhaps the best-known British aircraft ever built, the Second World War Spitfire fighter. And these too are links that tell of an integrated, cutting edge, and militarily outstanding descent from the past to the present. Other such genealogical stories are also possible, for instance, tracing lineage to the civil aircraft mentioned earlier, the Viscount. Here the link takes another form, pointing out that this aircraft had been built efficiently and to cost, and emphasizing that Vickers had a "track record of production management and on-time deliveries" (Gardner 1981, 31). And then again, reminding the reader that the Viscount had been designed, like all good civil aircraft, for ease of servicing and quick turnaround. Both points can also be read as an unkind cut, however, for in reading between the lines you are meant to understand that Vickers builds reliable and matter-of-fact aircraft and completes its projects to time, whereas English Electric does not. So this is the production of more genealogical similarities, similarities that make intercompany differences.

Small, versatile, easily serviced on a modular basis, deriving from the Spitfire and the Viscount, this is not a bad origin story. But consider this:

The 571 was a revolutionary proposal in that it offered the required blind terrain-following, nav-attack and weapons system as a fully integrated package—the complete opposite of the "add-on" afterwards school of thought. The argument was that the systems were the heart of the airplane and a high performance flying platform should be built around them. (Gardner 1981, 30)

"A revolutionary proposal." This is the historian of the British Aircraft Corporation, Charles Gardner, talking. My reason for drawing attention to this passage is that it makes another kind of cut, a division between the past and the present, between what are now being distinguished as "the 'add-on' afterwards school of thought" and the "fully integrated package." Gardner implies that we are witnessing a historical step change—and then he distributes value across that boundary in favor of whatever comes later and is thereby in touch with the present. It is the performance of a past where things were both different and not as good.[11]

Dis/continuities

So this is a world of cultural and historical discontinuities in the form of "revolutionary proposals." But take a look at this:

> Work was well advanced, and if it had been selected there is little doubt that the P17A could have been flying as a prototype by 1963. . . . It had another potential advantage in that it would have met the broad operational requirement, powered by a modified standard Bristol Siddeley Olympus engine. (Hastings 1966, 30)

This is Conservative MP Stephen Hastings, who wrote a book on the project. It's true that he's picking over the entrails after the event. He's reflecting on the fact that in the end the choice was for a "weapons system" aircraft of the kind preferred by Vickers Armstrong. And he's exploring the fact that for a variety of reasons (his book considers these in some depth) this aircraft ultimately came to a sticky end—his is a story that resonates in particular with those who worked for English Electric and their "evolutionary" P.17A. Most interesting here, however, is the way in which the polarities have been reversed. Here value is being distributed across the boundary the other way round, for the "weapons system revolution" is being performed as an obstacle rather than as a great leap forward. Continuity, accretion, descent, these would have been better. This is the lesson that we are being asked to draw.

So the distributive polarities may change. Differences between the present and the past may be desirable or they may not. In some stories about what came before, the past may fall from favor. In others it does

not.

But this is only half of the story.

An aircraft must be treated not merely as a flying machine but as a complete "weapons system". This phrase means the combination of airframe and engine, the armament needed to enable the aircraft to strike at its target, the radio by which the pilot is guided to action or home to base, the radar with which he locates his target and aims his weapons, and all the oxygen, cooling and other equipment which ensure the safety and efficiency of the crew. Since the failure of any one link could make a weapons system ineffective, the ideal would be that complete responsibility for co-ordinating the various components of the system should rest with one individual, the designer of the aircraft. Experience has shown that this is not completely attainable, but it is the intention to move in this direction as far as practical considerations allow. (HMSO 1955, 9)

The citation is from a government document, a statement of official policy that appeared in 1955. But now that I have cited it, it starts to make a difference to the ordering of the narrative. In particular, it makes a difference to the story told by Gardner, the story of revolution and discontinuity. Suddenly it looks as if Vickers's revolutionary proposal isn't so revolutionary after all. Another story is, or could be, performed—a story of continuity. This is a story of a different *kind* of continuity, one that tells of the links between government arms procurement policy and the 571 proposal made by Vickers, which in this new cultural context isn't revolutionary any more. Indeed, it isn't even entirely new.

So it is that we find ourselves back in a world of continuity, genealogy, and descent—albeit a continuity different in kind from that celebrated by English Electric.[12] And if I wanted to strengthen that narrative of descent I could tell stories, too, about the reasons for the government policy statement, about why the government came to favor a weapons-system approach. These stories would have to do with procurement policies in the United States and with certain unfortunate British projects which created aircraft that flew satisfactorily by themselves but when mounted with weapons turned out to have aerodynamic problems if they were fired.[13]

Back to a world of continuity.

Here's a proposition. Culture in all its forms—talk, technics, skill—is about *making and distributing similarities and differences,* about allocating them and re-allocating them. It is about trying to stabilize them or undermine them. It is about "the strategies [we] recognize and use and invent for making sense" (I'm citing Sharon Traweek again). But (this is the proposition) when we perform these allocations we also reflect, perform, instantiate, and form narratives that reflect and embody forms of *cultural bias.*[14]

Which we? That's a good question.[15] Let's just say for the moment, we who are the narrators, the cultural bricoleurs, the performers, the engineers or managers who write brochures, the historians who write company histories, the sociologists and technoscience students who tell stories about social interests and the social shaping of technology. We, all of us, have a *bias in favor of continuity,* narrative continuity—for instance (though not necessarily) in the form of descent or genealogy. This is one of the features of the tendency to perform singularity and the concomitant tendency to marginalize multiplicity. Which means that, at the same time, we tend to have a bias against discontinuity, against revolutions and step changes, and especially against multiplicity and that which cannot be assimilated.

So here is the suggestion: we tend to assume that if we cannot assimilate something, trace lines of similarity, explication, then somehow or other we have failed. Even if the events present themselves as discontinuous. Broken up. Or multiple.

Lethality, Cost, Size

To be sure, the cultural strategies of continuity come in various forms. Vickers Armstrong's way of making similarities and building connections wasn't like that of English Electric, not, at any rate, in certain important respects:

> Whilst unit cost has very considerable significance the really significant parameter is made up of cost/size/lethality. The aeroplane is designed to do a certain job—primarily strike—therefore the financial outlay per successful strike is the important thing, or in other words the cost of a given degree of lethality. The achieved lethality is bound up closely with vulnerability and vulnerability is closely bound up with size. Cost per pound of all up

weight is of no direct significance. It might, for instance, be possible to show that an aircraft of 45,000 a.u.w. had a higher cost £/lb. than an aircraft of 65,000 lb., but if they had the same range and speed characteristics, and navigation bombing systems of exactly equal capacity, the small aircraft would have greater lethality because it is less vulnerable due to its smaller size.

Therefore even if the unit cost of the two aircraft were the same (and in fact the smaller aircraft would be less) the small aeroplane is still cheaper because it offers more lethality per £ sterling. (Vickers Armstrong 1958b, 2–3)

This is a full-blown expression of the "weapons systems" approach, a performance thereof. It is one of the strategies of coordination of which I spoke in chapter 2. But this is an approach to cultural distribution that constitutes its objects by making connections not through time but across space. Let's call this virtual space, virtual because it is conceptual and contains such entities as cost, size, lethality and "lethality per £ sterling." Such is the storytelling mode preferred by Vickers. It connects with government policy statements but then performs conceptual worlds and novel connections even though the company would also be able to recount perfectly plausible narratives about descent and genealogy.

System Continuities

The culture of Vickers, or Vickers-Supermarine, is quite unlike that of English Electric. That would be a way of putting it, a good way of telling of the difference. We have here two design cultures that embody quite different traditions, very different skills, substantially different understandings about what goes with what, and quite different approaches to building and solving puzzles. Or, to put it in one of the languages used in the study of technoscience, we are dealing two different technological frames.[16] For the similarities made by English Electric take us to past exemplary achievements, while those of Vickers take us into . . . into what? Into "weapons systems." But what does this mean?

Connections not through time, but across a virtual space. In practice this becomes in part the question of the physical size of the aircraft. This is the issue at stake in the previous citation. But imagine that physical size is simply some kind of intermediary between killing and cost. If we think of it this way, then the narrative connections made by Vickers are more radical, more "abstract,"[17] than those of the government—radical or "abstract," that is, in the sense that the frame of the story no longer has to do with the aircraft or even with its specific components. For there has been a transmutation, and the aircraft has been turned into a *part* of the system, a role or even a set of roles that may be narrated together in one way or another, a role or set of roles that can be distributed and redistributed. So the story no longer has to do primarily with aircraft that have become a means to an end. Instead the narrative stabilities, the beginning and the end, the alpha and the omega, these have to do with killing and cost, with the *links* between killing and cost and the most efficient way of making those links.

This is a logic, a mode of distributing and making that we have discussed before. But let me also observe the following: there is a shift from diachronic to synchronic. Narrative similarities and differences trace themselves in a synchronic space rather than down lines of descent. The object is dissolved in this synchronic syntax for telling stories and making connections. I've also noted this too. For the focus is no longer the aircraft but rather the system, and the aircraft is being imagined as a product of that system or rather as a set of characteristics or as an expression of that system: size, number of engines, and (if one goes into it more carefully) radar cross-sections and training accidents. So what used to be an aircraft is turned into a set of places, roles, and features while the "aircraft itself," the form one sees when one looks at it, is shaped by these relations. Like the transmutation from work into text,[18] it is an object of variable geometry until the system stabilizes itself, an effect, an outcome. It is nothing in and of itself.[19]

Context. This suffers a distributional fate like that of the aircraft. For it too is desegregated and colonized. That is to say, to the extent it is important, it is brought within the syntax of the system as a set of places that may be told as having certain effects. This means that like the aircraft it isn't given, or if it is given then it is also malleable. Objects appear, but they take new forms: "lethality" instead of targets or people and their destruction. All of which is a feature of technology including, perhaps especially, war technology. Here "users" are being connected with the system, "configured" or shaped,[20] which means that inside and outside are not given in the order of things. For the system goes everywhere that is functionally relevant, and those objects that do exist are permeable, revisable.[21]

Finally this story forms a calculable space, a homogeneous space, within which to contain specific narrative threads. For there is no difference in kind between money and lethality. The two are interchangeable. Instead the issue is simply that of calculating the best exchange rate. Vickers has a technology for making these calculations and for drawing distinctions, distinctions, for instance, between large aircraft and small or between those companies that are able to make those calculations and those that are not.

If culture is a set of forms for distributing connections and disconnections that perform similarity and difference, for making narrative coordinations, then to work in terms of systems is to perform another cultural technology, a technology in addition to and distinct from that of genealogy. It is to perform a technology for making and distributing connections in which everything important is connected to everything else. It is to deploy a form of storytelling that tells of its universality, the generality of its calculus. In which everything important may be constituted, connected, and performed that way. And such is the genius of Vickers's proposal: it homogenizes, making all the story elements malleable, a function of everything else, a universal grammar.

It is no surprise that such an attractive cultural strategy for making similarity and difference should have found a ready place in the distributions of social science and in technoscience stories, this system of universal continuity.[22] It is not surprising that this strategy should narrate the differences between the human and the nonhuman as less than important—that they arise, as it were, after the event, as a function of the operation of the narrative distribution. It is also not surprising that such stories should reveal colonizing tendencies, unwillingnesses, or inabilities to deal with that which may not be assimilated.[23]

Size

Here is another story.

On the one hand, the Whitehall mandarins liked the English Electric design, the P.17A. It was aerodynamically excellent. But they weren't so sure about the capacity of English Electric to manage the project. On the other hand, the Vickers's 571 proposal was also good, especially in its commitment to systems thinking. And the management record of the firm was outstanding. Their conclusion was that "the right thing to do is . . . to give the task to the Vickers/English Electric combination, provided that the leadership is in the hands of Vickers and indeed in the person of Sir George Edwards" (AIR8/2196 1958a).

But this is simply a preface to the story I now want to make, a story to do with size, the size of the aircraft:

> It is desirable both from the point of view of development time and cost, that a proposed aircraft to any given specification should be as small as possible. For any project study the optimum size of aircraft is obtained by iteration during the initial design stages. The size of aircraft which emerges from this iteration process is a function of many variables. Wing area is determined by performance and aerodynamic requirements. Fuselage size is a function of engine size and the type of installation, volume of equipment, fuel and payload, aerodynamic stability requirements and the assumed percentage of the internal volume of the aircraft which can be utilised. (English Electric/Short Bros. 1958, 3.1.1)

These words from English Electric tell a story about the relationship between time, connection, and disconnection. But would another manufacturer have told a very different story? It seems unlikely:

> From the very beginning of our study of the G.O.R. we believed that if this project was to move forward into the realm of reality — or perhaps more aptly the realm of practical politics — it was essential that the cost of the whole project should be kept down to a minimum whilst fully meeting the requirement. This led us towards the small aircraft which, by concentrating the development effort on the equipment offers the most economic solution as well as showing advantages from a purely technical standpoint. (Vickers Armstrong 1958b, 2)

So these words from Vickers arrive at essentially the same conclusion, the conclusion that it is desirable to have a small aircraft. But though they both want small aircraft, the two companies are going to end up in very different places, at least for a time. Here is English Electric again:

> Abandonment of twin engines would be the only other way of achieving a smaller aircraft and this also involves a large reduction in the sortie pattern. This arrangement has not however been considered, due to the overwhelming pilot preference for a twin-engined arrangement even in the P.1B. (English Electric/Short Bros. 1958, 1.S.6)

This is unlike Vickers's preferred 571 proposal which was, as we have seen, for an aircraft with only one engine—and was, as a result, much smaller than the P.17A. So what was the difference? How did it come about? I have mentioned part of Vickers's reasoning, the equation of size, cost, and lethality that was related in the brochure to other interlinked stories about equipment miniaturization, integration, and space saving. But what of English Electric? Two engines, or so it reasoned, were better:

> This is because of the very high accident rate of supersonic aircraft following total engine failure, due to their very high rate of descent and the limitations of emergency power control systems. The argument for two engines in the present case is reinforced by the need to operate several times further from base than the P.1B and for a substantial time at low altitude where the glide capability would be much reduced. (English Electric/Short Bros. 1958, 1.S.6)

In other words, they are saying that a supersonic plane flies quite badly when it loses power—which is something that doesn't have much to recommend it. So it was that with two engines, the P.17A ended up with a design weight of 66,000 lbs. (English Electric/Short Bros. 1958, 1.1.4, Ransom and Fairclough 1987, 298).

It is possible, indeed easy, to link the difference in sizes to the stories about cultural difference discussed earlier. English Electric is telling a story about pilots, the experience of pilots who try to fly powerless planes. It is, to be sure, distinguishing between humans and nonhumans. So English Electric uses cultural genealogy to offer a reason for choosing two engines rather than one. Against this Vickers is making other similarities and differences, synchronic connections, systems connections that dissolve the distinction between human and nonhuman and lead to its own very different fixed points, those of cost and lethality. So these are narratives of social and technical shaping both, but they are made very differently.

It would be easy to do this, but I want to go somewhere else. I want to think about the mandarin response to this difference. For remember, by now we have the civil servants agreeing that they want the two firms to collaborate on an aircraft. There are going to be many problems, but one is quite simply that one of the designs is half as large as the other.

To jump forward, in the end the government will go for a larger version: "a study on the single versus twin engined aircraft was received 16th July, 1958. It showed fairly conclusively that the twin engined configuration is the less costly in accidents" (AIR8/2196 1958b). Vickers's calculations, their version of the relationship between cost, size, and lethality, these have been worn down, overturned, by alternative calculations made in government. That is a possible story. No doubt it was one that was performed in the corridors of Whitehall. But there are other possible connections too, other narratives. For instance, here's a second excerpt from the same document, referring to events that took place very slightly earlier, in June 1958: "The matter of a joint requirement for the Navy and Air Force still loomed large and Vickers Armstrong's submission to G.O.R.339 included a very promising single engine solution . . . it was decided *inter alia* that the Air Ministry would initiate a study to determine the economics and wisdom of having a single engine version of the aircraft" (AIR8/2196 1958b). This also tells about size, economics, and wisdom, but there are other actors too, new actors. Most notably, there is the navy. But what has the navy got to do with it? The story runs something like this:

The Navy has its own aircraft under development. Known as the NA 39 and later the Buccaneer, this is a small aircraft. It has to be small because it's intended to fly from aircraft carriers. But the NA 39 is also much slower than the GOR 339 aircraft and its range is much shorter. Though it is designed to drop nuclear weapons, its electronics and avionics are much less sophisticated than those being proposed for GOR 339.

Since March 1957 it has been government policy to press ahead with the development of ballistic missiles because Minister of Defence Duncan Sandys thinks that the age of the manned warplane is over. The future, or so he says, belongs to missiles, ballistic missiles that will drop hydrogen bombs on Moscow. And, sometime in the future, there will be antiballistic missiles that will meet and destroy the missiles that will rain down on London.

The Royal Air Force isn't persuaded by Sandys's vision. Indeed, most officers think that it is little more than science fiction, and even those who take strategic ballistic missiles seriously think that antiballistic missiles are a pipe dream. In addition, everyone believes that a whole lot of fighting isn't going to be possible with missiles of any kind. This is GOR 339 kind of work—for instance, surveillance, or tactical strikes on railways, factories, bridges, and armies. These are the kinds of bombing raids that would be needed if a war against the Russians didn't go nuclear, or at any rate strategically nuclear, in the first four minutes.

It has taken most of 1957 to persuade the minister of defence that this story might be the case and then to persuade him that, despite its apparent similarities, the Naval NA 39 is really quite different from the GOR 339 aircraft in speed, precision, and range. It has therefore taken most of 1957 to persuade him that it isn't a cheap option to stop the GOR 339 aircraft and simply to order the NA 39 for the air force as well as the navy.

By this point the outlines of a possible explanatory story have become clear, haven't they? Let's list the three main actors in this drama.

(1) Vickers wants a small GOR 339 aircraft for various reasons, but one is certainly that it thinks it will sell better.

(2) The Royal Air Force might, just might, be persuaded by Vickers's synchronic systems-derived arguments. But this is going to be uphill work, partly because pilots don't care for unpowered supersonic aircraft but also because if they accept the small plane, then this puts the air force position in jeopardy. This is because one of the differences between the NA 39 and GOR 339 falls, that of size. But if this happens then the Minister of Defence or the Treasury might force the RAF to buy an "improved" version of the NA 39.

(3) The navy doesn't want a small GOR 339, however desirable such

an aircraft might be in the abstract. This is quite simply because a plane in the hand is worth much more than two in the indeterminate future.

Interests

So this is a nice story, indeed a classic, built as it is around social interests, around the narrative trope that hidden, or more or less hidden, social interests shape decisions and outcomes.

They do it. That is, the people we study in social science tell their stories so. Often enough they distribute their realities that way. And *we* do it too, in our own social-science studies, in one form or another. We do it in analyses of the class- or gender-shaped character of technological change,[24] in social-interest theory, as elaborated by the "Edinburgh School,"[25] and in the studies of bureaucratic politics that come from political theory.[26] And then, though perhaps in a less structural form, we do it in the theory of *interessement* or translation found in actor-network theory.[27]

None of this is very surprising: there is no particular reason to imagine that our forms of cultural bias, the ways in which we make connections, would differ from those round about us. For this is simply another narrative strategy for creating, shuffling through, and assembling assorted bits and pieces, material, verbal, human; another way of finding connections and making similarities and differences, of finding connections that give some shape to the distribution and redistribution of specificities. This is a resource for storytelling available to anyone who is willing—and who is not?—to adopt the position that they see further, or better, *deeper* than those round about them.

The strategy of depth, of seeing deeper, may be related either to genealogy or to the synchronicity of system building. Interests may shape evolution, or they may be reflected in the structure of systems. There isn't too much to be said about this, one way or the other, except perhaps that interest stories aren't just smooth stories—though, like systems and genealogies, they certainly count as that. They aren't just smooth stories because they are also stories that manage to bring together oppositions—or, more precisely, they are stories that manage to bring together *apparent* inconsistencies. The strategy of seeing into the depths looks past dissimilarities, even clashes, by treating them as symptoms or superficialities and going beneath the surface to fundamental and coherent places that are said to be more real. That is, it welds what have been turned into superficialities into a single, more comprehensive narrative. Multiple cultures, conflicting cultures, similarities and differences, all may be subsumed to the potential of the narratives of depth, the strategies of storytelling in depth.

I said earlier that systems stories are totalizing and that they colonize. But now we find that the same is true for interest stories: these too are totalizing and colonizing. For opposition is understood and incorporated into the interest structure, the deep interest structure, that underlies and underpins appearances. Similarity is achieved in yet another guise—while multiplicity is displaced.

Cultural Bias

So what have we learned about the cultural distributions of military technoscience? We have learned that there is a bias against multiplicity or discontinuity, though not necessarily against difference when this can be subsumed *within* continuity. And we have learned that this performs itself in three great distributive forms, three variations in cultural strategy.

First, there is *genealogy,* which is a form of culture, of narrative, that makes its similarities and differences through time by performing lineages, lines of descent, generational or chronological origin stories. If we were to invent a social history of genealogy, perhaps we might say that this is an aristocratic form of storytelling belonging to and performing, in the first instance, the premodern.

Second, there is *system,* which effaces history and generates a synchronic and homogeneous space in which everything conceivable starts off by being similar in kind before being quantitatively distinguished and distributed. Thus, differences no longer exist in the generational order of things (as in genealogy) but are rather to be understood as the distributive consequences of contingent calculations. In an invented history of culture perhaps this would belong to the bourgeois era, being modern both in its incarnations as liberal polity and as market economy.[28]

Third, there is *interest.* Or, more generally, there is a cultural strategy that discerns realities (such as interests) that are hidden behind superficialities. It works by distributing entities into levels that are performed as qualitatively different. As a specificity, it may distribute narratives into two classes: the class of those narratives that are real (of which it is one) because they describe that place called reality which may be hidden from the common view; and the class of those narratives that are not real because they tell stories of realities that do not exist and even tend to mask reality. In a history of culture perhaps we'd want to locate this in critique and so, in critical modernity, as the antithesis of system.[29]

So there are three distributive forms, three ways of making connections that build worlds embodying different conditions of possibility, different ontological spaces. But though there are important differences between the three cultural strategies, there is also a similarity: a propensity to perform consistency, smoothness, and connection. As I noted earlier of genealogy, but now the point may be made more generally, each embodies a *cultural bias in favor of continuity.* Genealogy slips through generations. System, no doubt more radically, homogenizes everything that is, everything that could be told. And even interest or depth, which like genealogy makes differences in kind, does so within a tellable set of similarities.

Continuity over discontinuity, connection over disconnection, the effacing of multiplicity—such is the cultural bias of technoscience distributions and, no doubt, many other contemporary performances. Perhaps, one might venture, it is a feature of social science and technoscience, a modern grand narrative, a grand narrative that no doubt includes our own narratives, the distributions made by sociologists and technoscience students.[30] Narratives that, taken together, tend to perform themselves into being in other, nonverbal, material forms and therefore tend to make the narrative smoothnesses of technoscience, of modernity. And, to be sure, the smoothly centered subjects to which they correspond. Cultural bias works in favor of singularity, even though it also makes difference.

This effacement of multiplicity suggests that it might be useful to find ways of making culture that emphasize and perform discontinuity in addition to continuity; that are rough as well as smooth, stutter rather than, or as well as, speaking fluently; that perform patchworks alongside homogeneities and multiplicities alongside singularities; that imagine technoscience practice as a set of partial connections rather than as synoptic visions,[31] and run interference on

The Project

A project is a "plan, scheme; planned undertaking, especially by student(s) for presentation of results at a specified time" (*Concise Oxford Dictionary*). The term derives from the Latin *pro* (in front of, for, on behalf of, instead of, on account of) and *jacere* (to throw) (ibid.).

Perhaps, then, "the project" is the performance of a cultural bias in favor of continuity.

Perhaps "the project" is a performance of the three cultural orderings of continuity: genealogy, system, and interest or depth.

Perhaps it is possible because these three orderings, individually discovering their limits, are able to pass the baton of continuity to one another. In which case singularity is secured in a process of continual narrative shuffling that forever defers the interruptions and discontinuities. And the problem of multiplicity is effaced.

Effaced, but at the same time performed.

—The TSR2 project "itself."

—The project to study the TSR2.

—The project of technoscience studies.

—The project of technoscience.

—The modern project.

its grand narratives by refusing to come to the point. That perform another and different form of cultural bias.[32]

Postscript

It is sometimes said that to give up grand narrative is to embrace political conservatism. This is one of the criticisms made of postmodernism: it is simply about playing. It is not a coincidence, according to this argument, that at a moment in history when dominant groups find that they are under threat they suddenly discover the virtues of small narratives. This neoliberalism, or so it is said, is best understood as a form of neoconservatism, a new strategy for preserving existing distributions, an expression of the cultural logic of late capitalism.

No doubt there is something in such a complaint. Divide and rule was always an effective strategy of power. But the story can also be told quite differently. Is it the case that dominant groups are under threat? Is it the case that a strategy of partial connection amounts to the same thing as liberalism, neo- or otherwise?

If we opt for the discontinuities of stutter or the fractionalities of partiality, we will no doubt have to debate how dominances, asymmetries, and the uses of power might come to look that way. Both/and. Multiplicity/singularity. But this is precisely the point. One does not have to go all the way back to Louis Althusser to imagine that multiple cultural performances enact overlapping similarities and differences. One does not have to imagine that all the instruments are playing a single score to imagine that the conditions of possibility may shape themselves into asymmetries. One does not have to be able to tell the whole story from a single place to imagine that there may be asymmetries which perform themselves—yes—in distributed ways.

Perhaps, then, it is time to imagine multiplicity, fractionality, and partiality as alternative cultural strategies. And perhaps it is the moment to imagine the tools for apprehending distribution after homogeneity.

You don't have a map in your head, as a child. Later, you have the globe—the seas and the shapes—and you can't ever get back to that emptiness, that mystery. Knowing that there are other places, but not knowing where they are, or how to get there. —Penelope Lively, *City of the Mind*

Mimesis fuses brilliantly with alterity to achieve the connection necessary for magical effect, . . . a kind of electricity, an ac/dc pattern of rapid oscillations of difference. It is the artful combination, the playing with the combinatorial perplexity, that is necessary; a magnificent excessiveness over and beyond the fact that mimesis implies alterity as its flip-side. The full effect occurs when the necessary impossibility is attained, when mimesis becomes alterity. Then and only then can spirit and matter, history and nature, flow into each others' otherness. —Michael Taussig, *Mimesis and Alterity: A Particular History of the Senses*

The ground figure that emerges from the stories in the previous chapters is one of oscillation or displacement. On the one hand there is the normative simplicity of the modern project, which seeks to enact the god-eye and presupposes the ontological singularity of the world that it desires to know and make. This simplicity is sustained by the theory of perspectivalism that allows, indeed requires, different viewers to see different things when they look at an object. The hope, however, or the expectation, is that a single story may be told of an object that is equally singular. And on the other hand, there is the multiplicity of the so-called postmodern world, with its renunciation of grand narrative and its preference for an aesthetics of little stories. Modernist singularity and postmodernist multiplicity, the two stand in tension with one another.

This much is standard fare in the social sciences. Indeed, it is the customary terrain within which much of social-theory debate operates—and within which it is usual to stake out a position. To find, for instance, that "modernism" fails because it denies the lack of foundations that has precisely been generated by the restless machine of modernity as it dissolves all that is solid into air. Or because it leads to the barbarism of what Zygmunt Bauman calls "gardening" (Bauman 1989). Or, conversely, to find that "postmodernism" is a repudiation of intellectual and moral responsibility, and to assert that it is indeed possible to make claims about the world even if they turn out to be relatively provisional (see Giddens 1990).

But I am suggesting that there is a much more interesting way of looking at this oscillation, this tension between singularity and multiplicity—or between modernism and postmodernism. It is to note that *the two imply one another.* That for instance (to take the case discussed in the previous chapter) singularity precisely *sustains* itself by shifting endlessly between different stories—stories that are, themselves, singularities. But at the same time this means, as is obvious, that the fact of different singularities together also performs multiplicity. Together they are performing a more or less self-denying or self-effacing multiplicity—a deferral indeed of that which does not tell itself as singular in order to secure singularity.[1]

Looked at in this way, the reason the debates about "modernism" and "postmodernism" take the form that they do becomes clear. There is endless room for "postmodern" debunking of grand narrative— of forms of storytelling that announce themselves to be both comprehensive and (necessary concomitant) singular. Such debunking is easy because incompleteness and incoherence can always be found. And, conversely, there is endless scope for complaining precisely about that debunking because it denies the possibility of (real enough) singularity and the intellectual and moral commitment implied in taking a stand.

What I'm suggesting, however, is that it is much more interesting and productive to explore oscillation between certainties than it is to take a position in the debate. For that is what I am attempting to do: to explore metaphors for the processes of incompleteness that do not force us to one pole or the other and that do not, therefore, insist upon the fundamental character of (what has been turned into) a dualism. To echo Bruno Latour, the strategy I seek to articulate is neither "modernist" nor "postmodernist" in form—though unlike Latour I take it that the both/and logic applies not simply to humans and nonhumans on the one hand and hybrids on the other, but more generally to any entities (objects or subjects) that live in oscillation, which means, no doubt, all entities. So this is "nonmodernist" perhaps, though the term strikes me as unhappy because it also implies commitment to some kind of secular chronological ordering that it would be better to avoid.[2] And if there is room for postmodernism at all, this is only because it provides an attitude, or a set of techniques, that are initially helpful: in short because its skepticism secures the possibility that everything does not hold together, as is imagined in modernism, in singularity.

I am searching, then, for metaphors for thinking the oscillation between multiplicity and singularity, and for ways of reworking the narrative conditions of possibility performed in modern and postmodern storytelling. I also want to find ways of re-creating subjectivities that do not draw everything together but are not, conversely, simply fragmented and to explore the ways in which those oscillations perform themselves, their modalities, their modes of interference. That is the point of this chapter, in which I consider the slippages and deferrals that ground (without ever finally grounding) an aerodynamic formalism and seek to reinterpret the notion of "heterogeneity."

First Story

It was to be 84 feet long, 23 feet high, and 35 feet from wing tip to wing tip. And we have met it already. It was called the P.17A, and it was designed by English Electric in 1958 to replace the Canberra.

I will talk about the design of its wings. Like a paper dart, these were to be delta-shaped, their leading edge swept back at 50 degrees. They were to be thin, their thickness only 2.5 percent of their breadth at the tip. They were to be short and broad; their aspect ratio (the span from wing tip to wing tip divided by gross surface area) was to be 2.77. And their gross surface area was to be 597.3 square feet.[3]

So why were they to have this shape? An answer will take us to design, and to the *heterogeneity of design,* its distributions. To the patterns in its overlaps and interferences.

Look at the following. It comes from the English Electric brochure on the P.17A (English Electric/Short Bros. 1958, 2.1.9):

$$G = \frac{(\text{velocity} \times \text{lift slope})}{\text{wing loading}} = \frac{M \cdot a_t}{W/S} \qquad (1)$$

Let's define the terms, for these are terms that can be linked to the words that appear in the less formal part of the expression.

—M is Mach number, the speed of sound, so $M = 2$ would be twice the speed of sound, and so on.
—a_t is transonic lift slope; more about that in a moment.
—W is the weight of the aircraft.
—S is the wing area.
—And G is a measure of the response of the aircraft as it flies through vertical gusts of wind.

So the equation expresses what aerodynamicists call "gust response." It quantifies the susceptibility of an aircraft to vertical buffeting. The aircraft, or so the expression tells us, will be buffeted less if it weighs more, and it will be buffeted more if it flies faster, if it has a larger wing, and if its lift slope is higher.

Difference

The concern is with heterogeneity and its distributions, in centering and singularity, and the ways in which this is achieved in slippage. It is about overlap, difference, deferral, and singularity. It is about co-ordination and interference. The tools derive once again from semiotics. And indeed it is easy to apply semiotics to a formalism such as equation 1, for this is the distribution of a visible set of relations, a set of differences. And it is a set of differences that helps to determine the significance or role of the terms that are linked together. For instance, it establishes the difference between gust response and velocity. There are, as they say, "variables" that intervene between these, such as lift slope and wing loading. If everything else were equal, if these variables were not to intervene, then gust response and velocity would vary together—but they don't because it is rare for everything to be equal.

But is everything there? To pose the question is to suggest the answer. Something is missing. Indeed, much is missing. This is obvious. For the distributions made by formalisms don't stand alone. But what is missing?

Second Story

What happens if we magnify the formalism? It depends on what we choose to magnify and where we look. I'll magnify it in various ways in the course of this chapter, but let me start with the term left hanging earlier, that of lift slope. We already know something about lift slope. We know that it is related to, but different from, gust response and the other terms in the formalism. But outside the formalism, the term is idle. It is a short cut. It doesn't tell us anything. So what do we discover if we magnify it?

The answer is that it decomposes. It turns from a single term into a relation between two further terms. So this is another difference, another specified difference. And the new terms? These are *lift* and *angle of attack*.

Some definitions.

—Lift is the lifting force of a wing as it moves through the air. In engineering this is usually written C_L.

—Angle of attack, written a, is (roughly) the angle between the wing and the air through which it is traveling (figure 5.1).

FIGURE 5.1
Relationship between Angle of Attack and Lift. If the increase in lift with increase in angle of attack is small, then lift curve slope is low.

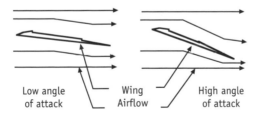

—And lift slope? This is the slope of the curve that links lift and speed for a given wing if they are laid out as the two coordinates of a graph (see figure 5.2).

All of this means that if lift slope is low then as the angle of attack alters, lift doesn't change much and the curve is flattish; if lift slope is high then lift changes more and the curve is steeper.

FIGURE 5.2
Curves Relating Lift
to Speed

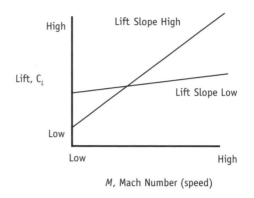

FIGURE 5.3

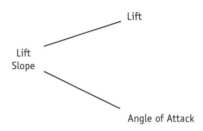

Heterogeneity/Simplicity

If we magnify the term "lift slope" in this way then we introduce a further set of differences. We might write them into equation (1) to produce something like this:

$$G = \frac{\text{velocity} \times (\text{change in lift coefficient}/ \text{change in angle of attack})}{\text{aircraft weight/wing area}} = \frac{M \cdot (dC_L/d\alpha)}{W/S} \quad (2)$$

We might work at this formalism to rearrange its terms and simplify it a little, but let's make another point. This new formalism is more complicated than the old one though it's not unmanageable, at least not yet. But if we were to expand the other terms—for instance unpacking the calculations that lead to Mach number, M—it would grow still further. And it could be expanded in other directions too.

What might we make of this? One answer is that design is all about distributing relations, relations of difference, but that only some of those relations are *relations of presence*. Only some of them crop up together on the page. And the corollary? This formalism performs many other relations. It also makes *relations of absence*. In one way or another, and for one reason or another, there are limits to the relations made present.

So what might the character of these absences be? The answer, or so I want to suggest, is that there are several *logics of absence* or alterity. I will point to some of these shortly. But, looking at the formalism above, don't we see a straightforward and immediate version of the logic of absence? This is the fact that it is easier to handle formalisms with fewer terms than those with more (though the same logic applies just as well to nonformalisms). So this, perhaps, is a basic design principle, a basic feature of the character of making simplicities — that *present complexity is self-limiting* (see Strathern 1991).

Perhaps we might call this *heterogeneity/simplicity.* Put the "heterogeneity" on one side for the moment: I'll come back to this shortly. By "simplicity" I mean, straightforwardly, that there is not enough room for everything. That not everything can crowd into a single place. That implosion, or perhaps better condensation, is impracticable. Not possible. Perhaps this is a general principle. But, linked to a concern with centering and singularity, it's what the actor-network theorists point to when they tell of "punctualization."[4] Complicated things come in simple packages; they are "black boxed" by such terms as *lift slope,* which can be used to make sense.

Third Story

The P.17A brochure continues. In the paragraph immediately after the formalism, we read the following: "By comparing several aircraft, of known characteristics, which have been flown in low altitude turbulence, it is possible to decide a maximum value for this parameter which will ensure a comfortable flight" (English Electric/Short Bros. 1958, 2.1.9). "This parameter" is *G.* Gust response again.

Heterogeneity and Absence/Presence

On the one hand the two excerpts are contiguous. It is reasonable to imagine continuity, co-presence, and more relations of difference — an organization in terms of the narrative relations discussed in chapter 2. But as we read on and a moment passes, so the field of presence starts to shift. Before, it was a matter of formalisms, terms that stood in quantifiable relations with one another. Now it is something different.

When we looked at that formalism we already knew that something was absent. We knew that there was one kind of logic at work, a logic of absence. We also knew that this absence was an engineering/algebraic logic, one of pragmatic simplicity, the business of limiting complexity in order to secure ease of manipulation. But there were other kinds of absence too. Indeed in order to make the narrative work, I let slip a clue. For by referring to "lift slope" as "idle" I traded on another absence: the suspicion that the reader would "know" what was meant by such terms as *weight* or *surface area*—which, by implication, were not idle. This, then, was the performance of another logic of absence.

The second excerpt takes us in another direction. For new kinds of relations are being performed, relations that no longer have to do with formalisms but rather with the flying of aircraft. I will delve into this shortly. But first let's focus on the changing relations of presence. The second excerpt performs a subtle shift. It "reminds" us what is *absent* from the formalism. But this is a double effect: it "reminds" us that there is no reference to "the real world," to what "actually happens" (as opposed to what might happen), but it also inserts that absent "real world" into the formalism, which means that after the second excerpt the real world is, as it were, both present and absent from the formalism, and that the formalism has started to acquire extra weight, an extra weight of difference. It has started to acquire this weight in the impossible interference between absence and presence.

This, then, is how I define heterogeneity, heterogeneity in design, and heterogeneity elsewhere. *Heterogeneity is an oscillation between absence and presence.* How it is that whatever is *not* there is *also* there. How that which *is* there is also *not* there. Both/and rather than either/or. Or both/and either/or and both/and. Heterogeneity, then, is about the differences that reside in connection and disconnection. Or, more precisely, it is about the ambivalent distributions entailed in dis/connection. Which means that simplicity is not simply about absence but it is also about presence. Hence the term *heterogeneity/ simplicity.*

Now we can ask: Are there other forms of absence/presence? Are there other heterogeneities?

Fourth Story

If we stay with the aircrew a little longer and search through the pile of documents we find this:

> The state of the pilots is variously described as "tired," "bathed in sweat," "weakness in limbs," "headache." The main factors causing fatigue appear to be several. There are oscillations in the higher frequencies to which various portions of the human anatomy respond . . . , moderate impacts which continually jar the pilot and throw him about, and occasional large gusts which frighten him by giving the aircraft a violent movement. In addition the pilot had the strain of carrying on with his job, and the worry whether the aircraft structure would stand up to the treatment. (English Electric 1957)

This paragraph is taken from an internal English Electric memorandum. Observing next that the pilots are "near the limit of their endurance," it continues:

> The navigator, who has his eyes on his instruments, will be more prone to sickness than the pilot who looks at the horizon. At the same time he will be trying to extract precise information from a variety of electronic equipment requiring fine adjustments to be made by hand. (English Electric 1957)

Heterogeneity/Materiality

Let's talk first of absence—for here we have a second form of absence. This isn't a matter of simplicity, or if it is, then it is a new form of simplicity. For this is *material absence*. Removed from the flat space occupied by the formalism, we find ourselves in the sweating world of the aircrew. We discover pilots who flew their creaking aircraft too low, who worried about whether the wings would break off, who were thrown about their cockpits, and who climbed shaking from their aircraft at the end of these flights.

If we are imaginative enough then perhaps we can smell the fear, feel the sweat on the bodies, taste the vomit. For this is another set of presences, another set of relations, another syntax. It is another set of differences—different presences that are absent from the space of algebra.

FIGURE 5.4

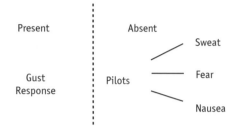

The corporeal or, if we include the aircraft, the corporeal with the technological—these are absent from the space of the page, from the formalism about *G,* gust response. This is the absence of materiality. For, in the way they write the P.17A brochure, there is no room for vomit. It does not fit. There is no room for sweat in formalisms. In the documents that are sent to the government ministries, there isn't enough space for the test aircraft they used, such as the Meteor aircraft. So they are removed, and not simply because there isn't enough room, but also because they are materials that do not perform themselves in the differences of the page, within a logic performed in algebraic difference.

And yet, these are absences that are also present. For *G* is there on the page. Gust response is fixed. It is fixed not by the other parameters that occur in the formalism (though these are fixed in their relations with one another). Rather, it is fixed in a set of relations, relations of absence/presence, to do with the suffering of the aircrew. "By comparing several aircraft, of known characteristics, which have been flown in low altitude turbulence," (I quote the sentence again) "it is possible to decide a maximum value for this parameter which will ensure a comfortable flight." A parameter to do with comfort, the comfort of a particular aircrew. Comfort that will allow them to perform the task of piloting the aircraft efficiently, properly.

Absence/presence, the absence of materiality that is also a presence—no doubt this is what those who write actor-network studies intend when they talk of "translation" and "chains of translation." (See, e.g., Latour 1993.) And this is a second oscillation in the distributions of heterogeneity: the absent presence of *materiality,* the Otherness of materials that don't fit in. But do.

Translation

Translation: the pun works best in the Romance languages: *traductore-tradittore, traduction-trahison,* to translate is to betray.

To translate is to connect, to displace, to move, to shift from one place, one modality, one form, to another while retaining something. Only something. Not everything. While therefore losing something. Betraying whatever is not carried over.

The only way to avoid betrayal is to mimic.

No. Even that is not good enough. To mimic is to move, to displace, through time.

The only way to avoid betrayal is to achieve identity. To be whatever it was, wherever it was, and whenever it was.

Such is the dream of communion, intellectual, ecstatic, religious, erotic.

Dreams have their place but this particular dream does not have much of a role to play in a world of oscillation between singularity and multiplicity, interference, and partial connection. Either that or, worse, it is a panacea.[5]

Fifth Story

Before we go on with this story of what is absent—about the absence, for instance, of fear—we need to go back to the formalism to understand what is happening to G and to forget, for the moment, the crew: "If the gust response parameter, G, is fixed to give a certain response level, and the operational Mach number and the aircraft weight are also fixed, then from (1) it is clear that $a_t \cdot S$ becomes constant." What is happening here? Let's deal with formalism first.

If G (gust response), M (speed), and W (weight) are fixed, then the only terms that still have freedom to move are a_t and S. It's easier to see what's going on if we rewrite the first expression

$$G = \frac{M \cdot a_t}{W/S} \tag{1}$$

as

$$G = \frac{M \cdot a_t \cdot S}{W} \tag{1.1}$$

But if G, S, and W are now fixed then equation (1.1) reveals that a_t multiplied by S, is (now going to be) a constant. When one goes up, the other goes down. It's a nice simplification: speed is inversely correlated to transonic lift slope.

So much for the formalism. But what of W and M, weight and speed? How come these have been fixed? Weight can wait. Let's take the case of speed. Look first at the previous page of the English Electric brochure. This tells us that "the essential design compromise implied by *O.R.339* is between high speed flight at low level, and operation from short airfields. The intermediate choice between a high-wing loading with a low aspect ratio to minimise gust response, and a large wing area assisted by high lift devices to provide plenty of lift at low speeds, must be resolved" (English Electric/Short Bros. 1958, 2.1.8).

Here there are a lot more complexities, but let's ignore most of them. Focus instead on the phrase "high speed at low level." So where has this come from? To answer we need to move to OR 339. We've come across this document already, so we know that it is an Air Ministry operational requirement.[6] It has been written by air force officers and tells a story about what a new aircraft is supposed to do. Part of paragraph 10 of OR 339 (Air Ministry 1958) runs as follows: "In order to minimise the effect of enemy defences, primary emphasis will be given to penetration to, and escape from, the target at low altitude." And part of paragraph 16 reads, "The penetration speed is to be in excess of $M = 0.9$ at sea level, with an ability to make a short burst at supersonic speed." So now we know why speed, M, is fixed. It is fixed "in order to minimise the effect of enemy defences." But let's push the paper chase one stage further. Let's ask, who is the "enemy"? And what are its "defences"?

Here is the opening paragraph of OR 339: "By 1965 a new aircraft will be required by the Royal Air Force for tactical strike and reconnaissance operations in limited war using nuclear and conventional weapons. Such an aircraft will enable the Royal Air Force to continue to make an effective contribution to the strength of SACEUR's shield forces, as well as to our other regional pacts." SACEUR is an acronym for Supreme Allied Commander Europe, which tells us, as if we didn't already know, that we have encountered another looming absence/presence: "We shall wish to consider whether there is a requirement for a low level weapon, either manned or unmanned, in case the Russian defences become effective against high flying aircraft and ballistic missiles" (AIR8/2167 1957). Here it is at last, made present not in OR 339 but in the correspondence of government ministers.

And the defenses of the Russian enemy? A background document to OR 339, referring to the earlier Canberra, states that "the Canberras, operated strictly at a low level, may continue to be effective until the enemy develops an efficient low level surface to air guided weapon" (AIR8/2014 1956). A defensive, surface-to-air, guided weapon. If the attacking plane is to evade such a weapon, it must fly at high speed and low altitude.

Heterogeneity/Otherness

This chain of distributive differences is complex, but we don't need to look into all of its ramifications. Retracing one line will do, one set of dis/connections.

Gust response, G, was fixed. It was fixed in a relation of materiality, material heterogeneity, the absence/presence of the sweating pilots. And Mach number, M, it turned out, was also fixed—because OR 339 said so. And why did OR 339 say so? In order to minimize the effect of enemy defenses. And the final set of dis/connections? The enemy turned out to be "the Russians" and the defenses "an efficient low level surface to air guided weapon." Which means that "fear" and "the Russians" are not simply outside the formalism *but also within it.*

None of this is empirically extraordinary. In tracing this chain we're not learning anything startling about the design of the P.17A. But we have learned something more about heterogeneity. We've learned that the enemy is within, that it is within the design, within the formalism. And the chain spells out one of the ways in which the enemy has been incorporated and assimilated.

This is another form of heterogeneity, another oscillation in differences that are both absent and present. For the enemy and its surface-to-air guided weapons are a part of the formalism, a part of the wing design, rigorously present. At the same time, like the extended formalism and the bodies of the pilots, they are just as rigorously absent. So this is a third form of heterogeneity, the *heterogeneity of tellable Otherness*. The enemy excluded, the foe that is necessary, necessarily included, necessarily a part of the center, necessarily Other.

"The Other": this is a *threat*. The air force officers who write operational requirements talk in just those terms. In their work they speak of "the threat." "The Russians and their surface to air guided weapons" are like Edward Said's Orientals (1991). They are necessary to the West, to its making of itself because they are dangerous, different, and antithetical. They play a similarly ambivalent role. For they are indeed a threat, a danger, something apart and something to be kept apart.

FIGURE 5.5

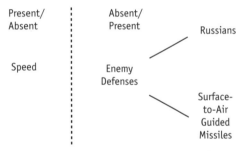

They deserve to be forbidden, excluded, kept at the periphery. Or, in the language of defense, they deserve "interdiction." So Otherness is a dangerous absence. But, at the same time, it is a promise, a seduction, a necessity, an incorporation, a need incorporated in its absence into the semiotics of presence. It is incorporated, for instance, into speed, *M,* and into the formalism linking gust response, *G,* to *M.* For without this incorporation, *M* might take any value. The wing of the P.17A might take a different shape. And the RAF's need for "a new aircraft"? Well, this too would look different, would disappear altogether.

Heterogeneity/Otherness. This is a third form of heterogeneity. It says that the forbidden, the abhorrent, sometimes even the unspeakable, is both present in and absent from whatever is being done, designed, or said.[8] Fear is distributed as an absent presence in the center, in the formalism.

Sixth Story

Let's go back to the fixing of parameters. Remember: "If the gust response parameter, G, is fixed to give a certain response level, and the operational Mach number and the aircraft weight are also fixed, then from (1) it is clear that $a_t \cdot S$ becomes constant." So G and M are fixed. Now let's turn to W. So why or how has weight been fixed? This is another paper chase. It takes us to a document that we have come across before:

It is desirable both from the point of view of development time and cost, that a proposed aircraft to any given specification should be as small as possible. For any project study the optimum size of aircraft is obtained by iteration during the initial

design stages. The size of aircraft which emerges from this iteration process is a function of many variables. Wing area is deter-
mined by performance and aerodynamic requirements. Fuselage
size is a function of engine size and the type of installation, volume of equipment, fuel and payload, aerodynamic stability requirements and the assumed percentages of the internal volume
of the aircraft which can be utilised. (English Electric/Short Bros.
1958, 2.1.8)

So there are many variables, too many for us to magnify. Let's stick
with engines. For aircraft size (and therefore weight) is not simply a
matter of the "size and type of installation." It's also, and even more
immediately, a function of the *number* of engines. Here is OR 339
again: "The Air Staff require the aircraft design to incorporate two engines" (Air Ministry 1958, para. 9). Two engines. But why? Well, we
already know the answer because we looked at the English Electric
brochure in the previous chapter. Pilots don't like flying supersonic
aircraft with only one engine when that engine fails.[9] So the pilots are
back again. This time they are not being frightened by oscillation or
nauseated, but they are worrying about something else. Another difference that is absent but present: the worry is that supersonic aircraft
are more likely to crash, and the OR 339 aircraft has to travel a long
way from home.

But there are other possible differences. We know that Vickers Armstrong wanted a single-engine aircraft: "From the very beginning of
our study of the G.O.R. we believed that if this project was to move
forward into the realm of reality—or perhaps more aptly the realm of
practical politics—it was essential that the cost of the whole project
should be kept down to a minimum whilst fully meeting the requirement. This led us towards the small aircraft which, by concentrating
the development effort on the equipment, offers the most economical
solution as well as showing advantages from a purely technical standpoint."[10] And these were the arguments: it would sell better; it would
be more lethal per pound spent; and it could interest the Royal Navy
because they might use it on their aircraft carriers (Vickers Armstrong
1958c, 2–3).

FIGURE 5.6

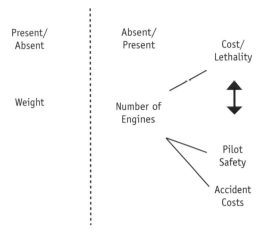

Heterogeneity/Noncoherence

Aircraft safety, pilot worry, the need to fly far from base. This set of considerations tends to fix W at a higher value and thus make the aircraft heavier. Cost, cost-effective lethality, naval use, practical politics, sales, this second set of considerations tends to fix W at a lower value and thus make the aircraft lighter.

So there are two sets of connections, two sets of relations of difference. This is old territory for those who study technoscience. It's a controversy. As we know, the Air Ministry is going to disagree with Vickers and stick with its large, twin-engine aircraft: "The reply by D.F.S. to D.O.R.(A)'s request for a study on the single versus twin engined aircraft was received 16th July. It showed fairly conclusively that the twin engined configuration is the less costly in accidents" (AIR8/2196 1958b, para. 43).

But if it is a controversy, it is something else too. It is another form of absence/presence. For controversy and disagreement are absent from W. They are absent from the formalism. There is no room for controversy in formalisms. Trade-offs, reciprocal relations, all kinds of subtle differences and distributions yes, but controversies no. *And noncoherences not at all.*

For, if the arguments about the size of the aircraft, about W, about the number of engines it should carry, are a form of controversy, they are also an expression of noncoherence, dispersal, and *lack* of connection. For the Air Ministry is talking about one thing while Vickers is talking about another: "We must be perfectly clear as to what is the principal objective of the design. It is to produce a tactical strike system for the use of the Royal Air Force in a limited war environment, or a 'warm peace' environment, and should thus be aimed at providing the maximum strike potential for a given amount of national effort. It is not—emphatically not in my view—to produce a vehicle to enable the Royal Air Force to carry out a given amount of peace-time flying for a minimum accident rate" (Vickers Armstrong 1958a, 1). Vickers is talking about cost/lethality, and the Air Ministry is talking about accident costs. This is a dialogue of the partially deaf. It is a dialogue in which the ministry decides—in which it "has" the power. But there is something else, a point to do with the absence/presence of noncoherence. For what is present encompasses, embodies, connects, makes links that are absent—except that such links aren't connections at all. They aren't connections because they aren't coherent and they aren't joined up into something consistent. Except that they are nevertheless brought together, in their noncoherence, in what is present. *(Present) coherence/(absent) noncoherence.* Like the performance of jokes in Freud's understanding, noncoherent distribution or interference *is a fifth version of heterogeneity.*

Seventh Story

Gust response, speed, weight, these are fixed. We are left with a_t, lift slope, the slope of the curve that tracks variations in lift against changes in angle of attack. We are left with this and the hope that its slope will be flat. But there is more. For instance, the stories are about transonic flight: How will the wing behave at roughly the speed of sound? And there are other questions; for example, how will it act at low speeds? So here's another complexity, one that I earlier chose to ignore. This is the quote again, from the English Electric brochure: "The essential design compromise implied by *O.R.339* is between high speed flight at low level, and operation from short airfields. The intermediate choice between a high-wing loading with a low aspect ratio to minimise gust response, and a large wing area assisted by high lift devices to provide plenty of lift at low speeds, must be resolved" (English Electric/Short Bros. 1958, 2.1.8). So gust response is important, but so too is take-off—which requires plenty of lift at low speeds. The brochure says:

Another convenient parameter is one which gives an indication of the relative response to gusts while achieving a given take-off distance. This may be expressed as P say, where

$$P = \left(\frac{a_t}{C_{LF}} \right) \qquad (3)$$

where C_{LF} is the maximum trimmed C_L, flaps down, in touch-down attitude. P must be a minimum for good design. (English Electric/Short Bros. 1958, 2.1.9)

We've met these terms before. A reminder:

—C_L is lift coefficient, roughly the lifting force of a wing: here, the lifting force of the wing as the plane comes into land with its flaps down.
—And a_t is lift curve slope, change in lift against change in angle of attack.

P therefore quantifies a hybrid relationship, the hope that it is possible to find a wing with low transonic gust response and high lift at landing.

But how to find a wing of the right shape? Of the right *planform*. This is a technical term and it is one of some importance. The brochure continues: "In the absence of comprehensive data on the effects of flaps on low aspect ratio wings, a comparison replacing C_{LF} by $C_{L\max}$ indicated that delta wings were superior to trapezoidal and swept wings" (English Electric/Short Bros. 1958, 2.1.9). The terms here?

—$C_{L\max}$ is the aerodynamicist's way of designating maximum lift.
—Low aspect ratio wings (a reminder) are wings that are short in relation to their area.
—Delta wings are triangular, like those of a paper dart.
—And a trapezoidal wing is shaped like a trapezium. That is, though the wing tip is parallel to the root of the wing, the leading and trailing edges converge toward that tip.

The paragraph then discusses planform:

Since it was thought possible that by using leading edge flaps on trapezoidal wings, higher values of C_{LF} might be obtained than those from delta wings, wind tunnel tests were carried out using a trapezoidal wing-body combination. In the event, these tests confirmed that the delta gave higher values of C_{LF}. The delta planform was also expected to have better transonic characteristics, and again high speed tests in our 18" tunnel on a family of aspect ratio = 2 planforms confirmed the unsatisfactory characteristics of trapezoidal wings, with sudden large aerodynamic centre movements at transonic speeds. This confirmed the choice of the delta planform. (English Electric/Short Bros. 1958, 2.1.9)

A further explanation. This time about aerodynamic center. As it moves through the air a wing lifts, but it does so by differing amounts in different parts of the wing. It's useful to simplify, however, and sum the effect of all these separate parts to create something called the aerodynamic center. Roughly, this is the place in the wing where the

FIGURE 5.7
Trapezoidal Wings

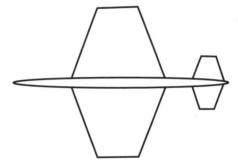

changes in overall lift occur as it flies faster or slower or its angle of attack changes. Above stalling speed the aerodynamic center doesn't shift much. At subsonic speeds it's about one quarter back from the leading edge for most wings. But at around the speed of sound the aerodynamic center tends to move backward. This isn't a disaster unless it moves quickly and jerkily, in which case the aircraft can be difficult to control—which would take us back to pilot sweat and fear.

So the English Electric engineers were looking at two things. One was aerodynamic center. Here the trapezoidal wing was a problem because the movement was "sudden" and "large." The delta wing was better. The second was $C_{L\max}$ (max, here, means maximum lift). Here

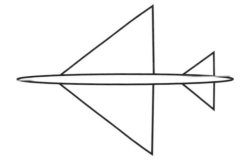

FIGURE 5.8
Delta Wings

there was a surprise: the delta wing was better again. On both counts
the trapezoidal wing came off worse.

Heterogeneity/Deferral

So there are two sets of relations: the link between planform, the
shape of the wing, and C_{LF}; and the link between planform and aerodynamic center. The delta wing is better—better, that is, in the wind
tunnel.

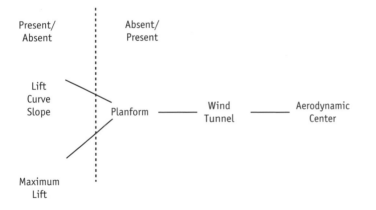

FIGURE 5.9

The wind tunnel is another instance of heterogeneity/materiality,
of distribution between absence and presence. On the one hand, there
are the flat surfaces of the drawing office that work to pull everything together, to center it; and on the other, there are the three-
dimensional models, materials, and measurements of the wind tunnel. So the wind tunnel is absent from the formalisms of the design
office and yet they are present too. But there is something more subtle

about the differences that emerge in that distribution. This is the fact that they are produced in movement, in a continuing process of displacement between materials and sites.

Perhaps one way of saying this is that it isn't possible to "sum up" the wing in the design office. The representation that appears in the design office, sets of formalisms, drawings, is incomplete, unfinished. It is not centered, it is not drawn together, because it *needs* the wind tunnel. It needs the differences that will be generated in the move to the wind tunnel. But the version of the wing that appears here is also incomplete and needs further attention, further attention by the design office, by stress engineers, machinists, metallurgists, and later by maintenance engineers and mechanics.

This is another oscillation of absence/presence. For the wing is present, all there, drawn out. But those lines also embody absence, the absent/presence of differences that are *deferred* and relations that are still to come. So the distributions here, the absent/presences are differences in movement. They involve displacement, displacement through time, in what Jacques Derrida calls *différance.* They involve an oscillatory distribution between the present/now and the absent/future. Or the absent/now and the present/future. In the heterogeneous interferences of time. In *heterogeneity/deferral.*[11]

Eighth Story

In English Electric's summary brochure there is a section at the beginning called "History." Here's part of the first paragraph: "Several widely-differing designs for a Canberra replacement aircraft were studied at Warton towards the end of 1956, and, by early 1957, calculations and wind tunnel tests had shown the optimum design to be an aircraft resembling the P.17 configuration. The merits of this configuration were confirmed by further tests, and the design was found to meet G.O.R. 339 requirements as these became known" (English Electric 1959). This paragraph is accompanied by three drawings of the P.17A that give an overall view of its geometry (see figure 5.10).

The full brochure offers a more abstract account: "The design process of a modern aircraft, especially a versatile one, could be summarised as obtaining the best combination of a large number of variables each one of which reacts on many of the others. The final product must meet each of its requirements roughly in proportion to the em-

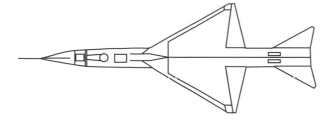

FIGURE 5.10
Plan of English
Electric P.17A

phasis placed on the relevant role" (English Electric/Short Bros. 1958, 2.1.8). This sentiment echoes those of the government White Paper on procurement that we have already come across:

> An aircraft must be treated not merely as a flying machine but as a complete "weapons system". This phrase means the combination of airframe and engine, the armament needed to enable the aircraft to strike at its target, the radio by which the pilot is guided to action or home to base, the radar with which he locates his target and aims his weapons, and all the oxygen, cooling and other equipment which ensure the safety and efficiency of the crew. Since the failure of any one link could make a weapons system ineffective, the ideal would be that complete responsibility for co-ordinating the various components of the system should rest with one individual, the designer of the aircraft. Experience has shown that this is not completely attainable, but it is the intention to move in this direction as far as practical considerations allow. (HMSO 1955, 9)

This was quoted in the previous chapter, but I'm citing it again now because I want to insert it into a different context.

The Architectures of Heterogeneity

I've been arguing that the formalisms of design are like other singularities—they are heterogeneous in character. That is, they work in the form of an oscillation between absence and presence, an oscillation that is one of the conditions of their possibility. This means that from the point of view of the center, the attempt to make singularity, the process is always ambivalent and incomplete. Viewed "technically," this ambivalence means there is always more to be done. The job is never finished; it is always an approximation. Looked at in that way, like the social theories of reflexive modernity, a formalism is, so to speak, a rule of thumb, something that might work for the time being—with any luck—but might need revising in the light of new events.

But that *is,* indeed, the "technical" way of looking at it, a way of imagining that attaches itself to the achievement of singularity while recognizing its necessary imperfections. Whereas if we embrace the logic of oscillation then formalism becomes something else. We need to say that it embodies, is the expression of, a set of tensions between what is present and what is absent but also present. Simplicity, materiality, Otherness, non/coherence, and deferral: these are some of the tensions and ambivalences enacted within the presence of a formalism. No doubt there are others.[12] And no doubt they are heterogeneous too, these distributions.

In this chapter I want to recover the ontological heterogeneity of this term, *heterogeneity.* I want to understand the tensions that are made in the processes of centering, in the desire to draw things together. This is difficult, itself a process full of tension. For when we talk of heterogeneity we also risk losing its oscillatory and unassimilable character: "I am arguing . . . that the stability and form of artifacts should be seen as a function of the interaction of heterogeneous elements as these are shaped and assimilated into a network" (Law 1987, 113).

This comes from an article that I wrote in which heterogeneity had to do with what I am now calling heterogeneity/materiality. My concern then was with system-building: the manipulation of all kinds of materials, technical and human. This is fine so far as it goes—though interestingly it re-echoes the desires of the anonymous authors of the 1955 government White Paper about weapons systems cited earlier. But by now it is clear that it needs to be nuanced or approached in another way. We need, or so I am suggesting, to avoid the flattening effect of imagining that there is, on the one hand, a great designer, a heterogeneous engineer, and on the other, a set of materially heterogeneous bits and pieces. Instead we need to hold onto the idea that the agent—the "actor" or the "actor-network"—is an agent, a center, a planner, a designer, only to the extent that matters are also decentered, unplanned, undesigned. To put it more strongly, we need to understand that to make a center is to generate and to be generated by a noncenter, a distribution of the conditions of possibility that is both present and not present.[13]

To efface this oscillation between singularity and multiplicity, to imagine heterogeneity simply from a control or engineering point of view is, then, another example of what I referred to in chapter 4 as cultural bias. For the notion of "heterogeneous engineering" may be understood in two ways. It may be treated as a way of thinking about oscillation, absence/presence, uncertainty, and the necessary Otherness that comes with the project of narrative centering. Alternatively, it may be used to describe and perform an *architecture of modernism* (Bauman 1989). No doubt there are different versions of this "modern project." No doubt they do different things. But, to put it too briefly, perhaps we might say that modernism is a way of being that seeks to improve the world, to engineer it, to build a better society. It does so by knowing, by gathering knowledge together, and then by deploying it in the attempt to order relations in the best possible way. It seeks to impose a specific and optimum distribution on its materials, human and otherwise.

The second version of "heterogeneous engineering" resonates with the benevolent and centering intention of this modernism. It catches something important about each of the "modernist" quotations cited earlier: the historical talk of the aircraft design and its "merits"; the "best combination of variables" cited in the English Electric statement of design philosophy; Vickers's systems talk with its trade-offs between cost and lethality; and the "combination" of elements mentioned in the government statement about weapons systems. In each it catches the utopian need to deal in different kinds of materials, technical and social, to center them, to handle them, to manage them. It does so with the characteristic modernist lack of concern with things in themselves—with, for instance, the distinction between human and nonhuman—for the perfect society involves both human *and* technical innovation (as we saw in the Vickers Armstrong design discussed in chapter 4). In each the second version of heterogeneous engineering catches the concern with simplification; with bringing materials together to optimize the outcome. It catches, that is, the need, the desire, to combine them together at a privileged place, that of the designer. In each it also catches the "semiotic" impulse that underpins the combination of somewhat pliable bits and pieces: the idea that components are a more or less malleable effect of a set of relations of difference; a set of relations that can be engineered to produce a better world. Perhaps, too, it also recognizes in each of these citations an acknowledgment of deferral, the deferral implied in the process of experiment, the trial and error, the iteration toward utopia.

The modernist version of heterogeneous engineering plays on all these notions. It resonates with them. But it misses the heterogeneities, those places that don't fit so well with the control impulse, that have forgotten that even the control impulse, the possibility of centering, is made by distribution into heterogeneity. Which means that it doesn't catch the heterogeneities of multiplicity. The "mess." The fact that things don't add up. The oscillations that make the mirage of the perfect center.

It would be good to reclaim heterogeneous distribution and its interferences from the flattening that comes with the modern project and to detach it from its utopianism. It would also be good to remove it from the concern to center, the concern to privilege a single place, the design/control place, the place of homogeneity, the place where whatever does not conform becomes a technical matter, an irritant, something to be managed, limited, and controlled. When it is recognized at all.

And instead? Well, instead it would be interesting to work with the idea that the conditions of possibility are lumpy and different, that they oscillate between singularity and multiplicity, taking fractional form and that heterogeneity is, indeed, heterogeneous. It would be interesting to imagine that absence/presence comes in indefinitely many forms and to investigate those forms. But most of all it would be good to imagine what might happen if the ambivalence of absence/presence were no longer a trouble, as something to be commanded and constrained, as distributions deserving of control from a single center. But instead to imagine their alterity as something to be welcomed and embraced.[14]

All media are mixed media, and all representations are heterogeneous; there are no "purely" visual or verbal arts, though the impulse to purify media is one of the central utopian gestures of modernism. —W. J. T. Mitchell, *Picture Theory*

The preceding chapters attend in one way or another to centers and how it is that the simplicities of centers are made. Objects, subjects, cultural continuities and formalisms—all are made smooth and centered, and in that making, all include, deny, and defer alterity, that which is not assimilable. The argument I am making, then, is that singularity arises out of that which is multiple. Or, more precisely, objects, subjects, and all the rest are never simple and singular but are also complex and multiple. Note that: are *also* complex and multiple. For the contrast between simplicity and complexity, between singularity and multiplicity, is not simply an either/or. Rather it is both a both/and *and* an either/or. Both single and multiple, the modernist logics of coherence (seek to) bury their noncoherences as they oscillate between one and many in the process described by Jacques Derrida as différance. For this process does not displace simplicity into a happy pluralism in which anything goes.[1] Rather, it is an attempt to come to terms with forms that perform themselves in ways that push us beyond what is easily told—beyond the limits set by the predominant conditions of possibility. Recognizing this oscillation enables us to come to terms with forms that are fractional—subjects and objects—more than one and less than many.

The oscillations between singularity and multiplicity set up their patterns of interference, patterns that take many forms. We have seen forms of interference that wrestle with the problem of multiplicity to produce relative singularity—the aircraft generated in the brochure discussed in chapter 2 was of that form.[2] We have seen interferences that produce immobility—such was the experience of multiple interpellation described in chapter 3, though that immobility turns out, or so I argue, to offer a valuable methodological lesson. We have seen the ways in which discontinuities in the social are effaced or deferred in the commitments to continuity embodied in the cultural bias described in chapter 4. And we have seen the various alterities that are both implied and displaced by the would-be self-present formalism described in chapter 5.

In this chapter I extend the study of oscillatory interference by revisiting the brochure. Again my concern is with the coherence of noncoherence. I explore the both/and *and* the either/or character of that brochure with respect to the complex interferences between its textual and pictorial contents. At the same time, I explore some of the ways complex gender distributions may be performed in a manner that extends both the object itself—the aircraft—and the performance of gender divisions themselves.[3]

Technics and Aesthetics

The brochure I started to explore in chapter 2 was published in 1962 by the British Aircraft Corporation and was intended for potential purchasers: that is, for a highly select group of military top brass, senior civil servants, and a few politicians in the UK, Australia, and one or two other possible purchasing countries. So it is already a special form of publication—a sales document. And, within that category, it is a special kind of sales document aimed at a small and more or less sophisticated readership. The cover and other pictorial ma-

EXHIBIT 6.1
Brochure Cover
(British Aircraft
Corporation 1962;
© Brooklands
Museum)

terial suggest that the brochure is also made to look "pretty," but only in some measure. So, to use a pair of terms that I hope I can in due course render more problematic, it exists in a place where what we tend to think of as the *technical* butts up against and interferes with what is often called the *aesthetic.*

So what might we make of this front cover? Some description and a brief recapitulation.

Most obviously it announces the title in big letters: "TSR2," and then, just below, in a smaller cursive typeface, it adds the words "Weapons System." In addition, toward the bottom, the name of the manufacturer, "British Aircraft Corporation," is highlighted in a sans serif type set against a darker background. And all this printing has, as a background, a perspectival but highly stylized depiction that may be understood as a view from the TSR2. For the reader of this brochure knows enough about the aircraft to know that one of its most important features is the ability to fly very fast at a height of only two hundred feet—and the view appears to be from two hundred feet.

So that is a description. But let me ask again: What should we make of this cover?[4]

Perhaps the most obvious response is to look at it briefly, to read the title, and then to turn the page. After all, it is just that, a *cover,* and as we know, covers announce what will follow. They attract us to the interesting and important contents—in this instance to the sixty pages of the brochure. And they frame it as separate, apart from that which follows, that which is not so set apart. So, even if we don't instantly move to the contents we might linger for no more than a moment on the design of the cover and say or think "very pretty," or "very stylized," or "very 1960s," or "I wonder what the air vice marshals thought of that?"

But to ask such questions is to hint at the possibility of treating it seriously as an object in its own right, as I started to do in chapter 2. In particular, I need to ask why on earth one would bother to spend time on something, yes a "detail," as seemingly unimportant as the front cover of a document when one might instead be studying its contents.

I offered an implicit answer to this question in chapter 2. There I talked about perspectivalism as a strategy for coordinating, a way of pasting together different object positions by assuming that they are complementary aspects of a singular object. Now, however, I want to explore this notion somewhat differently by restating a methodological prejudice. It may be that I am getting overly suspicious, but when something is said to be a "detail," I hear the performance of a fierce form of distribution (Law 2000c). More specifically, I hear the performance of a *hierarchical* distribution that (sometimes intentionally) demotes whole sets of phenomena to the subordinate class of minutiae, that which does not have to be taken seriously.[5] Which is what I suggest is happening here.[6]

So what of the aesthetic? What happens if we start to take it seriously? To put it briefly, as I earlier noted it is a trope of modernism to make purity, for instance, between the social and the technical, or between the textual (or even more specifically narrative) and the pictorial (or more specifically illustrative).[7] Though there are many reasons for supposing that the distinction does not work in this simple way, it is often suggested (at any rate by those who prefer text) that text is indeed prior and that it is narrative that counts, whereas illustration is just that, a supplement that "illustrates" what has already been told. A version of this reasoning may, however, be applied to the somewhat different distinction between the technical and the aesthetic. In this case the aesthetic is turned into an add-on, the packaging, so to speak, that covers, conceals, and misrepresents the real works hidden inside. It performs, that is, a version of the depth hermeneutics considered in chapter 4.

Not surprisingly, the values attaching to this division are contested. Thus the history of art, at least since Cézanne, may in part be understood as an attempt to create a space, a purely aesthetic space, that is valued precisely because it is removed from and rejects the functional.[8] Again, the "designers" of consumer technologies stormed into prominence in the early 1960s, inventing themselves by insisting that objects can be both functional *and* aesthetically appealing—or, indeed, that aesthetics is precisely an aspect of good design. But the existence of such struggles does not undermine the importance attached to performing a distribution between the functional and the aesthetic. On the contrary, it simply emphasizes its strength and persistence.[9]

So the character of the aesthetic is a site of struggle. But the TSR2 aircraft was not a consumer technology. It existed in that place, more or less foreign to the market, where the state liked, still likes, to pre-

Rationalization

To talk of rationalization is to play with ambiguity, a double entendre. Perhaps it is to talk of the way in which social life is rendered, has been rendered, more rational. Then again, perhaps it is to talk of justification, referring instead to something that is pasted on after the event.

Adrian Forty, a design historian, describes the way in which "London Transport" came into being in the 1920s and the 1930s. This organization grouped together the services of dozens of companies that previously ran the bus, tram, trolley-bus, and train services of the British capital (see Forty 1985, 222–38).

In the 1930s London Transport developed a distinctively modernist style for its new buses, trains, and underground stations. As a part of this, it also created a new typeface and a symbol—what is now called a logo—together with "uniforms" for its staff. The style of the buses and tube trains is particularly interesting. It took the form of flat and rounded surfaces. Like the cultural biases discussed earlier, these forms tended toward the smooth and avoided discontinuity. Little by little the London bus took on the characteristic appearance that it still in some measure has today. The angularities of the engineering of early twentieth-century coachwork were lost.

This design effort sought to create a new identity that would displace the identities and rivalries—as well as the dispersion—of the previous companies. But among the questions we might ask are the following. How deep did the design go? How far below the surface was it at work in its uniformity? Did it perform a new distinction between surface and depth? And what were the effects of replacing the visibly distributed stuttering that preceded it?

tend that it is playing market games while other logics are (also?) performing themselves. For here we are in the early 1960s, in the context of military technology. The readers of this brochure are powerful people with serious matters in hand to do with the defense of the realm, and they are required to make decisions about how to spend hundreds of millions of pounds. Which means (indeed this is a possible definition of "seriousness") that it is the pragmatics or the technics that matter through and through, whereas how things *appear* is quite another matter. It doesn't matter how the airplane looks, for in this logic looks are properly derived from function, and it matters even less how something like a brochure looks. The Military Sales Division of the British Aircraft Corporation does not share this view, however, or it would not have filled its brochure with pictures of the aircraft in various poses and commissioned this front cover, which is certainly noticeable, if not startling. All of which places the brochure itself in an interesting position, that of adding the aesthetic to a functional field where the distributions perform themselves rigorously, deeply, and without great apparent struggle.

So the question repeats itself: why attend to something as apparently secondary as the front cover of a military aircraft brochure? This question derives from the skepticism of a form of distribution between seriousness and lack thereof. It is performed, or so I'm suggesting, by the technologists and the decision makers and it also tends to perform itself, though possibly decreasingly, in many parts of social science. The aesthetic sometimes smuggles itself in a second-order way into our work, primarily if it is important to "the actors."

To be sure, it all depends on what we mean by "aesthetics," but here I want to press my methodological point. If we try to imagine ourselves back into the shoes of the air vice marshals of the 1960s and perform their distributions then (or so I guess) we'll notice the pretty or not-so-pretty pictures, but we will pass them by as we turn the pages with a small quantum of appreciation or perhaps a snort of irritation. In short, we will delegate them into the middle of next week and proceed to the serious business of reading the brochure, attending to its contents, attending to the narrative about the important matter of the military potential of the aircraft. *We will thus tend to reproduce the divisions made by those readers;* that is, we will reproduce the distinction between the technical and the aesthetic while effacing any of the questions that might have been asked if we had chosen to explore questions of "aesthetics." This, then, is one way of setting up the alternatives. The choice is whether or not to take seriously that which is "merely illustrative." Whether, in short, to go along with the demotion that little term "merely" implies.

Agency and Aesthetics

We have seen that the cover of the brochure is perspectival and that it is also about motion. It tells—or shows—that the TSR2 is capable of high-speed, low-altitude flight. Performed within a technical narrative, it is important to demonstrate this capability because it is a way of avoiding antiaircraft missiles. But, as I noted in chapter 2, this isn't the primary argument that's being made on the front cover, for here we are not in the world of technics. Instead, it presents a combination of perspectivalism, the Albertian apparatus of depth with the convention that we are looking through a window onto the world,[10] and a somewhat separate set of conventions about motion. Together they generate a fast-moving window and a series of contrasts that are relevant to, indeed perform, the distribution of agency. Let's look at that distribution a little more carefully. I want to suggest that it comes in three forms.

(1) *Active versus passive agency.* The eye moves. It moves toward the horizon, whereas the world stays in place. This much we have established. But this sense of motion is both strengthened and rendered more complex by other "details." Look at the ground. On the left there are trees, perhaps stylized cypresses or poplars. Then, in the middle we see what may be fields, which also seem to be a little like a passage, a passage that has opened up across the surface of the earth between the aircraft and that embracing vanishing point. And in the center at the bottom, and on the right, there are three or four buildings—houses or perhaps some of them are barns or light industrial buildings. We barely get to see any of this; instead what we see are outlines.

At this point the practical thought intrudes itself: that is simply how it is. For if you fly at 550 or 600 miles an hour at only two hundred feet above the ground then indeed you barely see what passes beneath you. This was no doubt part of the thinking of the artist, but the picture could have been drawn otherwise. For instance, we might, as in the television depictions of exemplary cruise-missile attacks in the Gulf War, have looked right into the windows of the buildings. But we don't, and we don't because to do so and to see such "detail" would undermine the dominant distribution that is here being performed: the division *between the dynamics of the aircraft and the statics of the world.*

We established that much in chapter 2. But if agents act, then they act because the capacity or propensity for action has been distributed in their direction. They have been constituted in that way. This was the claim with which I started. On the cover of the brochure we see the performance of such a distribution. The plane is being performed as an actor possessed of certain attributes, and in particular one capable of rapid and powerful movement, in the air. But that sense of movement and power is built in contrast for it cannot be separated from the buildings, the trees, and the fields. And the contrast works most strongly because of their pictured physical proximity, viewpoint, and ground. The relative motion is made visible by performing them as near and yet as almost infinitely far. Their physical proximity precisely marks a powerful boundary that performs them as static agents, agents made inert, close and yet Other, a division that is thereby reinforced by "the details."

Look again. Trees? Fields? Houses? To put it mildly, the combination is striking. What more conventional depictions of nature and culture could one ever hope to find? And if the buildings do not take the form of interiors (or even exteriors) by Johannes Vermeer, then this hardly matters because we already know that gardens, barns, and orchards are places of domestication, domesticity. That they are slow and soft, in some important senses passive. We know this, and the picture rests in part on this knowledge. But it also recursively performs it, performs them as passive, precisely because they are made to stand in contradistinction to the power of the aircraft, an aircraft made to inhabit a theater of activity, the theatre of work and war.[11] We are dealing not simply with a division between the aircraft and the houses; instead we are witnessing the (further) performance of a distribution between two great worlds: the world of the active and the world of

the passive. A distribution that implies their mutual dependence.

(2) *Transcendence versus mundanity.* Now move to the horizon. No, to beyond the horizon. The ability to make this move again works by combining Albertian perspectivalism and its promise of boundless (but not yet visible) volume—what Rotman (1987) calls a "zero point"—with the depiction of motion within that volume. The combination operates to generate a second contrast between that which is here, present, and that which is not. As with the formalism in the previous chapter, one might argue that the aircraft is both here and not here. It is present but also absent. It is here, for a split second, above these buildings. But it is also, already, for all intents and purposes, beyond the horizon at the dominant vanishing point.

So this is another version of absence/presence, of heterogeneity. But what lies beyond the horizon? Most obviously, the enemy Other (which is almost always implicit rather than explicit in this brochure). Perhaps, then, the contrast plays on the distinction discussed earlier, between activity and passivity, or domesticity and war, in which case the aircraft is shielding the home fires against the threat from beyond the horizon.[12] Such would be an iconographic distribution wholly consistent with the many strategic and technical narratives about the character of "the threat."

But there is something else as well, for we fly at the speed of sound toward the bright place of the vanishing point that is also the focal point of the picture. There is nothing much to detain us. We are pulled toward this point of attraction down the lines of movement. So it is a central place, the zero point, but it is also a bright place that seems to shine in our eyes like the rays of the setting sun.[13] And the symbolism is strong, for if it is the heart of the matter, a place to which we move, to which we are attracted, then it is also a place that holds out promise, the promise of illumination.

But what is the character of that illuminating promise? It might, of course, be the blinding "atomic flash" of nuclear detonation. But this is only one possibility. For if the aircraft is heading toward the enemy, then it is also heading toward the place where it will fulfill its mission, which means that it is hurrying to a dangerous place and a future that requires great courage. This need for courage generates a reversal, one in which invulnerability has given way to vulnerability, but a vulnerability that is now combined with skill in combat, the combat of machines and men, but also with bravery (for bravery is not possible without vulnerability). This combination makes that distant place extraordinary, a place of destiny, of desire, of consummation. For this, after all, is what the machine is all about: it is a machine to go and look, and then a machine to go and destroy.

So this is the second contrast: the performance of a distinction between what is here and what is not. A play around the heterogeneity of absence/presence that is also the performance of a division between the mundane and the everyday, between the rhopographic and the prosaic on the one hand, and the extraordinary, beautiful, megalographic and transcendental on the other.[14] The burst of light is therefore both literal and figurative. It is the light that takes the active agent and links him with the destiny of the hero. And the passive agent? Well, she has no place there, no place beyond. It is not for her to commune with the light, to test her skills in danger.

(3) *Invulnerability versus vulnerability.* I have mentioned vulnerability but I want to take this a little further by looking again at the relationship between the text of the front cover and the perspectival picture. If this is ambiguous, then perhaps there are two possibilities.

First, the picture may be read as background to the letters in the foreground. This would mean that the letters did not belong to the picture and its perspectivalism at all but rather hang in front of it like a flat screen or perhaps—like the glass at the top of an observation tower—a barrier that lets us see what is beyond without actually belonging to it. It also adds the warning that it is, indeed, a screen (for it is precisely the letters that make visible the otherwise invisible screen). If this is correct, the sense of proximate detachment is overwhelming. We (the aircraft? the reader?) are there, but we are also quite apart, in which case the cover performs a distinction between the invulnerable as against the vulnerability of the world.

Second, we may take the letters off the flat screen and insert them back into the picture. If we do this they turn themselves into an icon for the object that they also tell and can be located in two ways within that picture.[15] First, it may be that they are floating with their midpoint just above the line of the horizon, in which case (and contrary to my suggestion above) the picture is no longer "taken" from the aircraft but from immediately behind and below it at the split second that it flew past to meet its destiny at the vanishing point beyond the horizon. Second, it may be that the letters are instead written in giant proportions across the horizon, like the credits at the end of a Hollywood movie.[16] If this is right then the aircraft has been turned into a huge wall between us and the enemy, an invulnerable wall, which shields the vulnerable world that shelters behind it. Even so the trope is the same: invulnerability/vulnerability, invulnerability/dependence, it is another way of performing the same distribution of agency.[17]

Invulnerability/vulnerability. So this is a third distribution, indeed, a version of the classic distribution in which the domestic is shielded from (but provides for) work and war, which is also a distribution between the invulnerable eye and the vulnerable ground, a ground made all the more vulnerable by the way in which it has been frozen into stasis and made so that it could never touch the aircraft[18]— whereas, to be sure, the aircraft could touch the ground.[19]

Technical Struggle

I am suggesting that the front cover performs and distributes agency in three distinct and overlapping sets of contrasts. The overlap derives from the way in which they share the common propensity of distributing active agency in favor of the aircraft and away from features of its "context"—that is, from nature and culture. And the differences derive from the specificities of the various contrasts performed. Nature and culture are being made *passive.* They are being rendered *mundane.* And finally, they are being made *vulnerable.* By contrast, technology is by turns being made active, powerful, skillful, heroic, vulnerable, and invulnerable. No doubt there are various connections between these distributions; indeed the way in which the distributions interfere with one another is probably crucial. But it is also important that they are distinct. As we have seen, the possibility of different, somewhat incompatible, and coexisting distributions is not necessarily a problem at all. Rather their interferences may represent a source of strength rather than of weakness: where one distribution seems uncertain, there is tension, ambivalence, and displacement into another: singularity thereby grows out of noncoherent multiplicity.[20]

So there are three overlapping distributions. But if we now open the brochure and attend to its contents, we will find many more. Nature appears in exhibit 6.2, but this time the contrast between the technological and the natural is different: if the aircraft is still endowed with

The Established Disorder

"The possibility of different, somewhat incompatible, and coexisting distributions should not necessarily be seen as a problem. Rather their interferences may represent a source of strength rather than of weakness: where one distribution seems uncertain, there is displacement into another."

The established disorder: Donna Haraway's term (1994). Perhaps, then, a way of talking about the in/coherences of the oscillations between singularity and multiplicity.

The ethnographies of Sharon Traweek (1988a). Stories of intersections, bodies, interferences, making effects.

The empirical philosophy of Annemarie Mol (2002a). Performances of diseases which are/are not the same disease.

The minimalist music of Steve Reich. *The Desert Music.* Imperceptible shifts. Differences. Tensions.

The textured paintings of Adam Lowe. Layers upon layers, which interfere with one another to make recognitions.[21]

The novels of Paul Auster. Or, even more, *The Invention of Solitude* (Auster 1988). Overlapping and interacting pieces.

"Interlude V" in Bruno Latour's *Irreductions* (1988a), when "the white man" arrived with his disputing priests, merchants, civil servants, and technologies.

(Dis)ordering without a structure. Distributing without a scheme, without a plan.

The most difficult thing to talk about. Multiplicities. Partialities. The effects of interferences that are no-thing singular and yet make singularity.

agency, then this has started to change in character. Now it is "practical" or "technical," a performance or a distribution as a skill holder or perhaps as a puzzle solver, while nature has been distributed beyond the boundary of skill, into the realm of utilizable resources—which, to be sure, indexes and reproduces a large set of themes in Western and, in particular, Enlightenment thought.[22]

Here, then, we have removed ourselves from the realm of the "aesthetic" and been inserted back into the world of the "technical." For the surrounding passages explain how this "fixing" is done, albeit only in limited "detail" (remember that this is a sales brochure, not a technical manual). So there is difference but also similarity, one consistent similarity in which the narrative echoes the message of the front cover. The aircraft remains active, while nature is being rendered passive. For instance, in exhibit 6.3 the work, the intricate and

skillful work of the TSR2 is detailed. It acts in this way and that, while nature does nothing. Indeed, nature does so little that in this particular passage it has been effectively effaced. There is no need to detail the way in which landmarks don't move or how the features of the terrain reliably reflect the Doppler radar signals. The inactivity of nature, its stolid persistence, takes the form of routine, of a kind of brute durability, such that in the theater of aircraft action it is simple scenery. Or, to put it a little differently, the distribution rests on, presupposes, and *reproduces* realism.[23]

Activity/passivity: this similarity strengthens my original methodological conviction that it makes sense to treat the aesthetic and the technical symmetrically. For though the specifics of their distributions are different, they overlap and coordinate in ways that interfere with one another, and if we don't look at the distributions performed by nontechnical means, then we hobble ourselves. It is as if we were trying to reconstruct the sound of Mozart's *Requiem* while denying ourselves any knowledge of the woodwind.

But what of the *differences* between the contrasts? Much might be said of the distributions performed in exhibit 6.4. For instance, it

---

EXHIBIT 6.4 "The forward screen provides for head up display projection and withstands the hot air blast used for windscreen clearance, in addition to having adequate strength to withstand impact of a 1 lb. bird at transonic speeds." (British Aircraft Corporation 1962, 34)

---

simultaneously locates the proposed performance of the windscreen of the aircraft with respect to human, technical, and natural actors.[24] But perhaps it is the nonhuman actor—the intrusive presence of the bird—that is most interesting here, for the bird may be understood as an expression of nature. This means we're watching a further reworking of contrast between machine and nature; for, like the landmark, the bird is scarcely being performed as a resource. And if it is being made vulnerable, then this is not because it deserves protection like some endangered species. Instead the distribution is the other way round. The vulnerability of the bird is of no interest at all; it is rather an implicit *threat,* for no pilot, we understand, wants a bird hitting him in the face at high speed: shades of fear and sweating bodies. This suggests, all of a sudden, that nature is capable of action: the birds are fighting back.

In fact it is more complicated, though interestingly so. Two points. First, if the distribution of agency has shifted from aircraft to bird, then this is because of the extreme speed of the aircraft. In other words, it is precisely the performance of *technological* action that has also endowed *nature* with the capacity to act.[25] And second, it is the role of the windscreen to protect the aircraft and its pilot from the activity of these newly empowered birds, a distribution that is indeed effected in the text. So we are witnessing a process in which technology passes out agency with one hand while taking it away with the other, a trope, perhaps, of domestication. It allows the possibility of nontechnological action but only within the constraints of the technical and thereby sustains the active/passive distribution and, in particular, the performance of aircraft invulnerability.

This distributive mechanism recurs. For instance, in exhibit 6.5 nature (in the form of gusts) is again being performed as a potential threat by virtue of the actions of the aircraft (for the gusts would be no problem if one were standing still). But, in the same move, they are domesticated into a form of unproblematic passivity.

But what would happen if the bird weighed two pounds, a kilo? Or the gusts became extreme? Wouldn't this imply a redistribution of agency? This question is always elided in the brochure, though it is, to be sure, endlessly implied. Indeed, given both its contents and what we know of air-force thinking and practice in the early 1960s, one way, a very important and relevant way, of reading the brochure is precisely as an exercise in threat containment. For why else would the distributions endlessly perform the mastery, the agency, and the invulnerability of the aircraft? The answer is straightforward: this invulnerability is quite uncertain and the methods that it performs to secure whatever invulnerability it has achieved are (as they say) at or beyond the cutting edge of technology, which is another way of saying that they are iffy, open to doubt, and might break down, rendering the aircraft vulnerable.[26]

But if the vulnerability is not admitted directly, if it is distributed to nature—or to culture in the form of the enemy—then how is it admitted indirectly? The treatment of the birds and the gusts establishes the paradigm procedure. Vulnerability is admitted indirectly by performing the agency of the Other only if it can be countermanded and agency can be redistributed back to the aircraft. This is visible in exhibits 6.6, 6.7, and 6.8. The potency of agents other than the aircraft

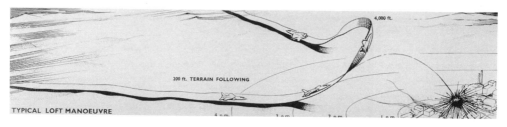

TYPICAL LOFT MANOEUVRE

EXHIBIT 6.7
"Typical Loft
Manoeuvre" (British
Aircraft Corporation
1962, 13;
© Brooklands
Museum)

is implied in exhibits 6.6 and 6.7 without being explicitly performed. Exhibit 6.6—and certainly this would have been understood by any likely reader of the brochure in 1962—implies that at medium altitudes the aircraft is vulnerable to surface-to-air, radar-guided missiles but that this isn't the case at low altitudes (at which point the desired distribution of agency is reaffirmed). And, as we know, the aircraft flies low, which means that invulnerability is performed yet again.

But there is a tension, for if exhibit 6.6 performs invulnerability, then exhibit 6.7 depicts a typical method for nuclear bombing, a "loft manoeuvre." In this maneuver the bomb is lobbed in an arc from some distance away onto the target so that the aircraft can escape destruction in the subsequent explosion. To do this the aircraft is flown upward at a predetermined speed and angle and the bomb is released at a precisely calculated moment. The picture, exhibit 6.7, shows this. The maneuver is tricky, but it works to distribute potency in favor of the aircraft and vulnerability to the target. Such is the effect of exhibit 6.7, in which we observe a conventional depiction of a large explosion in a factory—while the aircraft has turned away and is already making good its escape.

So there is a distribution that makes aircraft invulnerability and target vulnerability. On the other hand, the loft maneuver contains and domesticates one possible self-induced vulnerability (here, nuclear

EXHIBIT 6.8 "The serious exposure of the aircraft to ground fire which occurs when release angles of more than 90 degrees are used can be avoided by using a 'button hook' manoeuvre. In this attack the pilot flies over the target and returns to conduct a shallow loft attack from another direction. Vulnerability studies indicate that this is preferable to the use of over-the-shoulder loft attacks." (British Aircraft Corporation 1962, 13) self-destruction) but induces a second,[27] for when we look at exhibit 6.7 we discover that at only two miles from the target the aircraft reaches an altitude of four thousand feet. This may not be a "medium altitude," but is certainly more than the recommended two hundred feet and so reorganizes the vulnerability/invulnerability equation yet again such that *both* the aircraft *and* its target are destroyed. This certainly poses a serious problem for the economy of vulnerability, and the problem is treated overtly in exhibit 6.8, one of the few passages in the brochure that distributes military vulnerability quite explicitly back to the aircraft. But (here is my proposition) *it does so because the risk of vulnerability can be contained* and defused. For in this passage the "button hook" maneuver plays the role of windscreen and the distribution accordingly, once again, contains the threat.

Leakages, Redistributions

Technics is about that which is "serious." This is the first characterization, the first distribution, that has allowed us to distinguish between that which is performed as technical and that which is performed as merely aesthetic. But my argument has been that the struggling distributions of technics need all the help they can get, especially in the context of a state-of-the-art military technology. Technics are greatly assisted by the mobilization of aesthetic reinforcements and their further distributions of agency and passivity in favor of technology. One might add that the border disputes between the social and the technical (for instance, among sociologists of science and technology) are precisely about the distribution of seriousness.[28]

So technics is about "seriousness." But (and here we revisit a form of heterogeneity we have already discussed) technics is also about deferral. Such would be a second possible characterization of technology. Not, of course, deferral "in general," for this is a chronic condition, but rather *deferral away from technical agency*. For within technology the sky, as they say, is the limit. Anything is possible. Anything should be possible. Within the pragmatics of technology; limits are acknowledged only if they can be immediately redistributed back to the Other, the Other of nature or the Other of culture. Or, perhaps most of all in the optimism of the Enlightenment, to the Other of the future (see exhibit 6.9) in the form of the projectile that throws itself ahead, ordering the present by turning it into the means of the future.

This is what we've seen for the TSR2, though this movement, one might observe, is the general form of deferral within which the research and development industry has created its profitable niche.

Seriousness, the deferral of limits, and the denial of agency—this brochure distributes active agency away from nature and culture in multiple ways. For instance, there is one moment (I think only one) when the distribution seems redolent of rape (exhibit 6.10).[29] Much

more common is the use of the anodyne, a smooth and matter-of-fact language of practicality or impersonality. For instance, the term "delivery" (exhibit 6.11) seems to connect as much with the routines

of the postal service as with the broken bodies of warfare.[30] Again, phrases such as "optimum weapon performance" work to locate bombing within the realm of the balance sheet, the calculation of costs and benefits. And finally, its various drawings and sketches visualize the depicted explosions as destroying toylike buildings or vehicles in which, one might add, people do not appear (see exhibit 6.12).

Well, this is standard albeit lethally important stuff. Most of those who criticize the industry of modern warfare complain of the dehumanizing character of military talk. And, to be sure, the enemy is

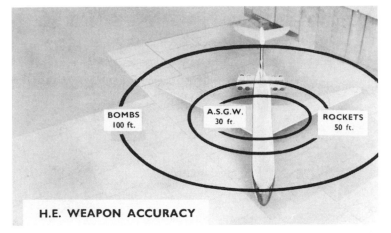

H.E. WEAPON ACCURACY

BOMBS
100 ft.

A.S.G.W.
30 ft.

ROCKETS
50 ft.

indeed being technicized here, that is, constituted as set of practical sources of possible resistance that can be overcome or contained like the birds hitting the windscreen. (Human) agency is therefore being distributed away from the enemy, which is turned into a set of objects; indeed the role of human agency throughout is being minimized. So the enemy has no capacity for successful or important action in general, and the effects of the "punishment" it receives are impersonal rather than personal.

Posed in this particular way this argument is theoretically humanist, but the argument that I want to make doesn't draw on the wellsprings of humanist orderings. I am keen to avoid assuming that certain distributions—for instance between humans and nonhumans—are given in the order of things.[31] The distributions that we're witnessing are more complicated and ambivalent, with contrasts that shift, human and nonhuman, object and subject, with complex relations and occasional reversals of polarities. Thus if human beings are complex and heterogeneous, it turns out that technologies are similarly complex distributive effects, complex and heterogeneous. They are *made* technical, made technical in a variety of different ways, but they also overlap, leak across the boundaries, and end up being performed, albeit in less obvious ways, as partaking of nature or culture, as heroic or childlike.

Heroism we have seen. This was performed by the front cover of the brochure. But exhibits 6.13 and 6.14 are pictures of the aircraft on the ground and they accompany text that describes how the TSR2 may be

Production and Consumption

Svetlana Alpers (1988; 1989) describes what she calls "Rembrandt's enterprise." That is, she describes the way in which Rembrandt set about producing his paintings in a manner at once technical and painterly, social (as in the organization of his studio) and economic (helping to create a market that recognized and performed genius).

Mieke Bal (1991) explores the paintings of Rembrandt, the semiotics of their surfaces. Particularly interested in the complexities of gendering, she uses resources from poststructuralism and particularly Lacanian psychology to detect or impute gender-related fears and complexities to "Rembrandt's" paintings.

This is one version, a specific version, of the difference between a concern with "production" on the one hand and "consumption" on the other. One version of the difference would run something like this: concern with "production" tends to produce potentially converging stories that perform context, genealogy, or interests, whereas concern with "consumption," at least in many of its incarnations, is more likely to produce diverging stories as interpretation is piled upon interpretation.

Is this divergence, indeed this multiplication of stories, to be treated as a problem? I asked myself this when I talked to audiences about the pictorial material of the brochure and invited participation and feedback. Collectively, we showed a remarkable capacity to make more and more "interpretations." Talking around visual materials seems to have this effect. Indeed, perhaps we might to add this to the list of differences between the "technical" and the "aesthetic": at least in this version the "aesthetic" *invites* interpretation whereas the "technical" does not. But what should we make of the proliferation of interpretations?

The austere view insists that it is proper to stay close to the material: that there is, as it were, no quality control when interpretations multiply in this way—a position that was forcibly expressed to me on a number of occasions. An alternative view is that something interesting is going on. This is that (cultural) products are complex and multiple in their effects, and that this—and the interferences that go with it— is a phenomenon in its own right. It may be, then, that this divergence is a feature of the performance of the "established (dis)order," of the chronic oscillation between multiplicity and singularity.

We have come across the latter view at a number of points in the earlier chapters of this book. Here it is, expressed again: "The possibility of different, somewhat incompatible, and coexisting distributions should not necessarily be seen as a problem. Rather their interferences may represent a source of strength rather than of weakness: where one distribution seems uncertain, there is displacement into another."

flown from dispersed airstrips. So the narrative distribution is again about vulnerability and invulnerability. Though it is not explored in the brochure, the narrative implies that in the event of nuclear war, large air bases will be destroyed in the first few minutes of conflict. Thus one of the virtues of the TSR2 is that it can operate independently of such bases for considerable periods of time, taking off from short runways or rough airstrips. And, with some specially designed additional equipment (including the vehicle that is also featured in exhibits 6.13 and 6.14), it can refuel and maintain itself.

This, then, is the point of these two drawings. They illustrate an argument about independence, a version of the colonization described in chapter 2, and therefore, like the front cover and much of the text, they help to perform the machine as invulnerable. But the drawings perform that invulnerability in an interesting way, by juxtaposing the aircraft with nature in a manner quite unlike that of the front cover.

RAPID REACTION STANDBY

The drawings do this partly because the static/dynamic distinction has disappeared. The aircraft is at rest; indeed, it is inactive. And (this is the point) it has made itself invulnerable precisely by virtue of this inactivity. So it hides in the bushes that therefore act to conceal it. Nature is no longer reduced to frozen immobility but has been turned into an active agent in its own right. It is acting as a shield in a manner not unlike the TSR2 in the iconography of the front cover. A reversal has occurred.

This, then, is a "technical" way of narrating the redistribution. But other things are also happening. In particular, iconographically and distributively, the contrast with earlier narrative and other pictorial representations in the brochure could hardly be more striking. (Compare exhibit 6.15.) For, yes, the movement, the power, these have disappeared. But what should we make of those bushes? Of the contrast between the bucolic scenes surrounding the aircraft and the aircraft itself? And of the pipes and people that populate the scene?

One answer is that in their conventional naturalism they offer a realist guarantee to an argument, the pragmatic argument, being made in the text. For these depictions are quite unlike the front cover in style. Though certainly conventional, they represent a range of everyday objects: trees, bushes, meadows, a lowering—perhaps menacing—sky. These are recognizable representations, representations that juxtapose themselves (perhaps rather startlingly, but no doubt that is the point) with the aircraft and its apparatus. But the juxtaposition, and in particular the pictorial realism of the depiction of nature, suggests that it is entirely realistic to imagine what might otherwise

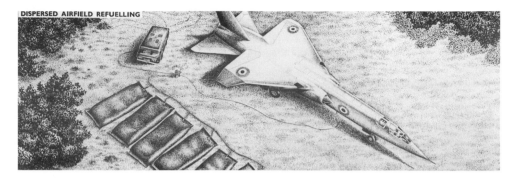

DISPERSED AIRFIELD REFUELLING

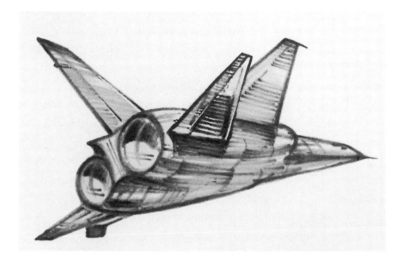

be unimaginable: the aircraft under such bucolic circumstances. The drawings may therefore be imagined as a kind of guarantee, a further warranty for the possibility of dispersal. The promise of refueling in the backwoods—this is a promise that can be fulfilled.[32]

But there is something else. The front cover performed contrasts between the active and the passive, the transcendental and the mundano, and between the invulnerable and the vulnerable. In each case the ground, nature, and culture, were distributed into the latter half of these pairs. But now there seems to have been a radical change. In the new drawings the technological, the aircraft, starts to partake of the features of nature and culture, to partake of passivity, mundanity, and vulnerability as opposed to excitement, heroism, and action in the air.

So it is that we see a surround of soft meadows, trees, and bushes. For by drawing a gentle landscape it becomes a place of rest and nurture, with all the tropes that this carries. For instance, there is husbandry. What is it, one might ask, that grows in this particular garden? What fruits does it bear? Is it dragons' teeth? For what grows is a weapon, a weapon of war or, more abstractly, a potential, a potential for action. Thus the aircraft is something that grows, grows quietly in potential and (it is understood) its quiescence is merely a stage, a moment—as will be revealed when it leaves the garden and that potential is unleashed.

So husbandry is one possibility but *domesticity* is another, perhaps related, gender trope. For on the front cover the aircraft was removed from the domestic. Indeed, we never saw it pictorially represented at all. But here, now, it returns for a time when there is need (as the technicians might put it) for resupply. But this distribution, which may indeed be proposed in technical terms, butts up against and interferes with the tropes of domesticity that both complement and escape it— tropes that have, for instance, to do with nurturing. We thus appreciate, without it having been said in as many words, that any man (or is it boy? or is it technology?) needs to return home at the end of a hard day's work and war to be sheltered, enveloped, protected. To be cared for, at any rate, by a putatively female figure.

Except that there are further complexities. Look at the human figures in exhibit 6.13. Those that we see appear to be men, men with pipes and equipment. Yes, we can tell that they are technicians. That is the story that the practical narratives of the technical will recount. But other contrasts and divisions are also being made, divisions that have to do with the distribution of gender. For, or so it appears, (parts of?) nature are being rendered male.[33] Or, perhaps better, nature is being displaced yet again, as if, for instance, instead of going home at the end of the day's work, the aircraft were rather being sent to the garage. Or, more pointedly, to the intensive care unit with its array of monitors, cables, bags of fluids, and drips as it enacts an intense and interventionary agency—that treats the body precisely as a (failing) system.[34]

But there is yet another possibility that has to do with gendering. This says that in certain instances patriarchal culture endows that which is *female* with potency. One of these—and one of the most visible—is in the performance of bodies. It may be found, for instance, in those distributions which constitute woman as a dangerous form of sexuality, dangerous that is, to man, for instance as the *femme fatale,* the source of forbidden knowledge, of forbidden power, a danger. In the iconography, this is a source of power constituted alongside and uncertainly held at bay by that other great patriarchal myth, woman as innocent, gentle, innocuous. All in all, as a source of virtue, either bland and passive, or active but in the form of nurturing.[35] In exhibit 6.13 we are not dealing with gendered or sexualized human bodies. But even so, it seems that something similar is going on; that is, we see an analogous alternation between innocence and blandness on the one hand, and dangerous physical potency on the other. An alternation in which blandness and one-dimensionality seek, with only partial success, to efface the dangers of materiality.

Blandness we have seen. It is everywhere in the bowdlerized and anodyne language of the technical. But it is also in the language of systems that permeates the brochure as well as other aspects of the project. We have seen that this is a colonizing language that says everything is normal, everything is under control, and everything is being monitored, calculated, and corrected. So there is the ubiquitous performance of the anodyne. But not very far away, albeit partially effaced, there is also the performance of danger. For the aircraft is a dangerous tool. It is, to be sure, a danger to the enemy. It is dangerous to those who fly it, testing as it does the limits of their heroism and skill. And finally it is dangerous to us all since its nuclear use would no doubt lead to consequences that would extinguish everyone. So these pictures show it sitting on the ground hiding from the threat, but they also show it embodying a risk to those who surround it, to those homelands it is precisely intended to shield. For this is a machine with potential, the potential for destruction, not only of the enemy but also of the power that wields it.

In which case the distributive dynamics become explicable. For now it appears that the superficial blandness is no longer simply a technical matter. Rather it is a device that, to a most limited degree, hides the knowledge of the dangerous and self-destructive potency of a weapon that would, if ever used, wreak havoc not only on its intended targets but also on those who deployed it. This hidden knowledge is what makes it possible to argue that this aircraft is also performed as if it were a woman, a dangerous mistress, in a distribution that connects through complexity with some of the performances of patriarchal sexual difference.[36]

Perhaps, then, the aircraft is ambivalent in terms of gendering, or even in terms of sexuality. Perhaps it is gendered or sexed, sometimes performed as one thing, sometimes as another. If this is right then the distribution of potency, of agency, is sustained in part by such abrupt redistributions of gendering. By their interferences with one another. And so it is that the "aesthetic," out of place in the "technical," reappears. It reappears in the form of the "illustrations" that, by demoting themselves to the status of "illustrations," are precisely able to propose distributions that strengthen the shifting performance of the aircraft as agent. But they are able to do so in a way that excites no comment, by other and technically outrageous means.

Ending

If agency is a matter of multiple distributions, and those distributions are effected in many different coordinations and contrasts, then to understand the character of agency it becomes necessary to explore the character of these interferences. It becomes important to explore the ways in which they overlap or don't. And it also becomes important to explore the Others that are generated in the course of setting up those contrasts—which means, predominantly, the distributions of passivity and their corresponding interferences.

The specific conclusion I want to draw in this chapter is that the performance of technological agency is complex. Perhaps technologies are indeed predominantly characterized by their capacity for action in a series of partially related contrasts. In the present case we have come across a series of pairs: activity and passivity, invulnerability and vulnerability, transcendence and mundanity, resource-using and constitution as a resource, containment and contained threat, rapist and victim, technical controller and technical outcome, man and woman, and woman and man. All of these are different (in some cases considerably so), but they all tend to perform the technology as an active and controlling agent and to distribute those contrasts in a way that simultaneously performs the passivity and vulnerability of (aspects of) nature and culture.

I am arguing, then, that the distribution being built by the brochure

Interpretation

One way of insisting on the rigor of "interpretation" is to work with materials that appear side by side. One would look, for instance, to references about gender in the materials under study before making any argument about gendering.

There are no explicit references to gendering in the TSR2 brochure. The austere view is that this renders it impossible to talk of gendering in this context. An alternative view is that this requirement is unnecessarily restricting: that if we allow only those categories which are to be found in the materials to form part of the analysis then we are limiting the ways in which we may juxtapose materials to generate effects.

The austere response in turn to this is that if we are allowed to juxtapose, for instance, a text by Griselda Pollock (1988) on gendering in pre-Raphaelite and impressionist art with the illustrations in a mid-twentieth-century aircraft brochure, then the effects are of a juxtaposition that *we* have made, which might therefore be made in quite different ways by others: that we are, in short, in danger of seeing what we want to see.

In the context of distributions such as gendering this is an ever-present difficulty: that, so to speak, we know what we are going to find before we start looking. Hey presto, there it is again! The countervailing pull is, however, the sense, which is simultaneously a blessing and a curse, that gendering is indeed endlessly performed.

There is no way to resolve this tension. But it is possible to make it more complex by refusing to imagine that gendering is a single distribution, or indeed a limited set of distributions and allocations, but rather is embodied and performed, as Stefan Hirschauer, Annemarie Mol (Hirschauer and Mol 1995), and Ingunn Moser (Law and Moser 1999) suggest, not as gender in the singular but rather in endlessly complex and partially connected genderings and genderings-relevant performances (all ungrammatically but deliberately put in the plural).

is strong precisely because of its complexity. It is a complex of interferences between different and partially connected strategies for distributing agency and passivity. For if the distributions are different, indeed so different that in places they appear to be in direct contradiction, those contradictions are problematic only if they are brought together. But this does not happen. Often they work to reinforce one another, but because they are distributed, kept apart, even their inconsistencies are not necessarily troubling. In a multiple or a fractional world constituted from partial connections,[37] there is no great premium on "overall" consistency—for there *is* no overall viewpoint, no god-eye view. Thus the division between the "purified" realms of the aesthetic and the "strictly technical" is precisely a source of potential power. For if these are said to have nothing to do with one another, then they can work in quite contrary ways to generate mutually supporting distributions or singularities. This helps to explain why the aesthetic might be so deeply removed from and subordinated to the technical in modernism and yet also remain so important to it. For the aesthetic entertains contrasts that are impossible within the technical or the pragmatic, contrasts that are discursively inconsistent or outrageous but that often enough work *with* those of the technical to produce singularity.

So agency, including technical agency, is performed in both technical and aesthetic distributions. But if these strategies for coordination have their specificities (and of course it is consistent with the argument that this is the case) we need to study both.[38] It becomes important to avoid treating the aesthetic as "merely illustrative" while attending to what is taken to be "serious," for to do so is to set draconian and quite unnecessary restrictions on our understanding of the distributions made by technical agency and its Others. But the converse is also true: we also need to attend to that which calls itself "technical." It is important to avoid restrictions that perform purity while, all around us, we are being distributed by impurities in ways that simply pretend to purity.

In short, once again, we are witnesses to the operation of an oscillation between singularity and multiplicity, an oscillation that juxtaposes multiplicities in a pattern of interferences that are the necessary condition for the strength of singularity thereby generated.

Soyez réaliste! Demandez l'impossible!—Text of French student poster at the time of the 1968 French évenements

S'il y a une tradition qui nous singularise, c'est, me semble-t-il, celle qui a nom "politique." La question de savoir ce qu'est la cité, qui lui appartient, quels droits, quelles responsabilités traduisent cette appartenance, et les mouvements de lutte, inventant des exigences, des obligations et des identités nouvelles, transformant les modes de'appartenance, sont ce qui singularise d'abord notre histoire.—Isabelle Stenders, "Le Médecin et le Charlatan," in Tobie Nathan and Isabelle Stengers, *Médecins et Sorciers*

Introduction
So there are multiplicities, multiple subjects and multiple objects. There are interacting performances. And there are many distributions between subjects and objects. Or simply between different objects. And then there are overlaps, resonances, alignments, coordinations, and interferences. Performances, multiplicities, distributions, and interferences come together as a package: it is not possible to talk about one without, at least by implication, talking of the others. And the singularities of "the modern project" arise from the interferences between multiplicities produced in that characteristic oscillation between one and many. Singularities arise even if the interferences and the multiplicities are effaced in deferral—are rarely visible, except perhaps as a technical problem to be addressed and resolved in the next bout of problem solving, the next attempt to render "truly singular."

In this chapter I explore the coordinating interferences of political decision making. I'm interested in capital *P* Politics, the kind of politics that calls itself "Politics," that makes itself big and important, the methods by which it assures its distributions. And at the same time I am concerned with the overlaps and interferences implied in a notion like "decision making." My case is the 1965 cancellation of the TSR2 project. Behind schedule and much more expensive than anticipated, cancellation, long debated, became a real possibility with the election of a Labour government in late 1964—a government that replaced thirteen years of Conservative administration.

EXHIBIT 7.1 "The discussion showed there had been a certain divergence amongst those concerned. James Callaghan, as Chancellor of the Exchequer, wanted to cancel the plane altogether for purely financial reasons. Ranged against him were (a) Denis Healey, who wanted to cancel the TSR2 and to substitute the American F-111-A, which would mean a certain saving of money but an enormous increase of outlay in dollars; and (b) Roy Jenkins, who wanted to cancel the TSR2 and replace it with a British plane—which was roughly George Brown's view as well; and (c) George Wigg, who held the view that we might have to cancel both but we musn't make any decision until we had finished the strategic reappraisal which would show what kind of plane was required." (Crossman 1975, 190–91)

Decision

Exhibit 7.1 is a record, one of several, of the cabinet meeting that led to the cancellation of the TSR2. It is drawn from the diaries of one of the participants, Cabinet Minister Richard Crossman (1975). Crossman was Minister for Housing in the government at the time of the decision to cancel—though perhaps in saying this I have already given too much away. But never mind. Let's attend to Crossman and what he writes. So what should we make of this?

No doubt there are many possibilities, but first I want to note that this is another distribution. We know the semiotic version of this argument: objects are being made, realities brought putatively into being. It is a performance and not simply a description. So Crossman generates, inter alia, four locations, places, or possibilities, each with its own specific attributes. He achieves this in various ways, but most straightforwardly, he does so by simply arraying them in the form of a list. This maneuver tends, as I shall shortly argue, to perform their equivalence for certain purposes.

What, then, of the specifics of Crossman's list? It is very important to say that his array is not idiosyncratic, some kind of invention unique to its author. For if narratives are indeed performative, then it is important to ask how much, to what extent, where, and how, the distributions that they entail are being performed, which implies a series of questions about interference that demand empirically complex answers. These answers will have us both attending to the differences between distributions just as much as their similarities and referring to material forms quite other than talk.[1] I shall touch on both of these issues in what follows. But in the present context we can simply say that Crossman's division performs, or at any rate assists in the performance of, a version of a distribution that is also being enacted in a wide variety of other locations—and indeed, by a wide variety of participants and observers. It is, in other words, not some strange aberration.

Let me give some examples of other related lists or arrays.

—Though, since it comes from the same source, this is not the strongest form of evidence, a little later in Crossman's own narrative we come across another, similar, distribution (exhibit 7.2).[2]
—Harold Wilson, who was Prime Minister and chaired the cabinet meeting, performs his own array that roughly coincides with the first two (exhibit 7.3).
—Though exhibit 7.4 does not reproduce the list in full, observers close to Denis Healey, the Defence Minister, describe a choice that maps on to those performed by Wilson and Crossman (exhibit 7.4).
—An extract from a Ministry of Defence press statement released on the day the TSR2 cancellation was announced generates a further distribution that can, again, be related to Crossman's (exhibit 7.5).

So it is important that these lists perform themselves in ways that tend to overlap—though we will need to attend carefully to the ways

---

EXHIBIT 7.2 "In the end, after another confusing discussion, Harold Wilson summed up: there were three possibilities. Possibility 1 was to cancel TSR2 without taking up the American option. Possibility 2 was to cancel while taking up the option. And possibility 3 was to keep TSR2 for the time being and make our final decision after we had finished the strategic reappraisal." (Crossman 1975, 191)

EXHIBIT 7.3 "But we had to have a decision, and the Cabinet was called again for 10.00 p.m. By midnight I had to resolve a difficult . . . decision. The Cabinet was split three ways; some favoured continuing with TSR 2; some favoured its outright cancellation; and the third group supported the Defence Secretary's view that TSR 2 should go but that its military role should be taken over by an order for American Phantoms, together with one for a number of F 111As." (Wilson 1971, 89–90)

EXHIBIT 7.4 "The conclusion that TSR-2 was expendable was made possible because a low-cost substitute existed in the high performance multi-mission F-111, which the American government was prepared to sell to Britain. Had this aircraft not been available then TSR-2 might have been saved. One of Healey's top planners admitted that 'the F-111 made cancellation of TSR-2 possible.'" (Reed and Williams 1971, 183)

EXHIBIT 7.5 "It will not be possible to define . . . [operational] tasks precisely until the defence review is completed later this year. This review may show that the number of aircraft required with TSR.2 performance characteristics may be substantially below the existing TSR.2 programme. On certain hypotheses about long term commitments it might even be possible to re-shape our defences in such a way as to dispense with this type of aircraft altogether. We shall make every effort to see how far existing or planned British aircraft . . . will meet the whole or part of the requirement. In order to ensure that our Services have appropriate aircraft in sufficient numbers H.M.G. have secured an option from the United States Government on the F.111A aircraft at a price per aircraft which even on a full scale programme would represent less than half the estimated total TSR.2 research, development and production cost." (*Defence* 1965, para. 5)

they fail to overlap in due course. It is also important that they overlap, or don't, in a number of different locations—though here, to be sure, I have considered only linguistic distributions while the performance (or otherwise) of dispersal across space and time in alternative materials is equally relevant.

Nevertheless, it is significant that various aircraft—the TSR2 itself, the F111, the F111A, the Phantom, and (here unnamed) British warplanes—keep on reappearing. It's going to be significant because what we usually think of as decision making—and here Political decision making—may be understood as the performance of certain forms of overlapping distribution. These distributions resemble one another or may at any rate (and the nuance is vital) be *made* to resemble one another. Though multiple, they share, at least in some measure, certain strategic features, features that help to render them also as singular.[3]

So what, then, are those strategic features? What are the "technologies" of decisions, or Political decisions—that make themselves important? What kinds of distributions do these attempt? These are the questions that I now want to tackle.

The Politics of Decisions 1. Reality and the Disappearance of Fantasy

The first distribution is more than a distribution. It is another of these great dualisms, the performance of a great divide between reality and fantasy. None of the exhibits actually says anything about this, presumably because there is no need to. But look, nonetheless, at the way this is done. For instance, all the exhibits take it for granted that the possibilities on offer are *mutually exclusive,* that they are indeed just that, "options"—which means that decision makers need to make a choice between possible scenarios, with the possibility of one, but only one, future reality. Thus the need for "hard choices" is performed for, and by, the British cabinet,[4] and the possibility of what the poststructuralists sometimes call "undecidability" disappears. Or, if it doesn't disappear, it is at least severely circumscribed and treated as a "technical" matter to be dealt with by (temporary) postponement, in the form of deferral that I have already discussed.[5] Pursuing more than one option is thus performed as a fantasy.

But even before the four "options" are brought into being, a distribution has already taken place, one that frames the list, reduces it. This is a distribution between *that which is possible, and that which is not.* For it is perfectly possible to imagine other possibilities, in principle. One might imagine, for instance, keeping TSR2 and buying an American aircraft, or doing away with the whole lot, TSR2 and any of its alternatives, or abandoning NATO, joining the Warsaw Pact and buying Russian aircraft, or, for that matter, abandoning any form of military defense at all. Such options are not inconceivable. But by the time the decision is being considered, these and any other options have been removed from the universe of possibilities or, perhaps more likely, were never conceived as options in the first place. *They have thus been performed as imaginary rather than real.*[6]

Fantasy and reality. They say of politics that it is the art of the possible, a place where "hard decisions" are made. But this itself is a performative distribution. It is performed in each of the exhibits cited earlier. It is an interference, an overlap. Or it is a coordination. If I were being aggressive I would add that it is also self-serving because it works to distinguish between so-called "dreamers" and "realists"—in favor of the latter, to be sure, who are thus built up as hard-headed heroes. Perhaps, however, it would be better to say more evenhandedly that a commitment to the importance of taking "hard decisions"—in Politics as elsewhere—is the art of enforcing the very distinction between reality and fantasy and of insisting on the division as one of the foundations of things. This division, for instance, confines fantasy to fairy tales or the dreamier realms of the academy and reality to the world, and then, to be sure, allocates specific possible futures between these two classes.[7]

So this is an ontological performance—the particular definition of the conditions of possibility that frames and also enacts decision making. For even in performances that make quite different specific allocations to reality and fantasy, that disagree about the reality or otherwise of the possibilities being debated, *the great division between reality and fantasy is being performed* and sustained—collaboratively, so to speak.

Debate in High Politics, this performance of the art of the possible, thus turns around boundary disputes: about what might be classed as real and what might not, but never about the existence of the boundary itself, or, indeed, the existence of these two great regions.[8] So this is the first great distribution, the first coordination, the first great technology of decision making. It is the abolition of the space that exists between fantasy and reality and the abolition of the possibility of living in that space.[9]

The Politics of Decisions 2. The Disappearing "Political"

The second overlapping distribution effaces certain forms of being, and then, more importantly, effaces the fact that they have been effaced. This, to be sure, is something that has concerned all those who ever wrote about ideology or the one-dimensionality of the political. But the difference is this: I have no particular notion about *what* is being effaced and I want to make the argument empirically. This journey will take us first into an inquiry about what is *not* being effaced and how it is distributed. So how does this work?

The answer is that it varies. It may come in discursive, mathematical, tabular, or pictorial form: any of these may generate one form or another of a list. But if we stick for now with the discursive, then in the present case each of the exhibits performs relations between a series of options. Each performs relations that distribute these options within the class of realities, distribute them as more or less desirable. It may do so directly as, for instance, in exhibit 7.5, or indirectly and by implication, as in exhibit 7.6.

EXHIBIT 7.6 "The decision has been taken after a thorough review of all the information that can be made available. The basic facts are that the TSR.2 is too expensive and has got to be stopped. The planned programme for the TSR.2 would have cost about £750m. for research, development and production." (*Defence* 1965, para. 2)

So such discursive moves operate to rank options. But this is just the beginning, for we are not dealing with a single discourse, a single mode of distribution. Rather, options are being mounted, performed, and ranked in several ways and in several discourses: indeed, within a multidiscursive space. For instance, the last part of exhibit 7.5, and exhibit 7.6 (again taken from the Ministry of Defence press release) both talk of costs. They say that the F111A is cheaper than the TSR2. But the earlier part of exhibit 7.5 argues on quite different, strategic grounds—hinting that under certain circumstances *both* the TSR2 *and* the F111A might compare unfavorably with alternative British aircraft[10] (a possibility obliquely picked up by Crossman in exhibit 7.7).

EXHIBIT 7.7 "We are cutting back the British aircraft industry in order to concentrate on maintaining our imperial position East of Suez. And we are doing that not because we need these bases ourselves but because the Americans can't defend the Far East on their own and need us there." (Crossman 1975, 156)

And there are further kinds of discursive distributions, for instance, to do with the viability of the British aircraft industry or the national balance of payments (see exhibit 7.1), but since I am concerned with the similarities and the interferences, we do not need to go into these here. For if each is mobilized to perform difference, to construct and distribute aircraft options, and to *rank* those options, then this all depends on the very *possibility* of comparison. Each difference depends on making judgments, judgments between options, judgments that depend on their similarity. It thus involves the performance of a series of homogenizing moves. This may sound odd, given that we are dealing with one of the most controversial decisions in British defense policy since the Second World War—or, indeed, that we are dealing with a multidiscursive or at least a fractionally discursive rather than a monodiscursive space (a point to which I shall return). Nonetheless, the performance of these differences is framed within the *possibility* of accountability. It depends on, it could never be mustered without, the performance of a framing of similarity, of singularity.[11]

Two points.

(1) Decision making tends to perform itself as the cockpit of difference. It is where, as it were, different options are brought together and focused. Nowhere is this clearer than in High Politics, where the differences that are said to be important are worked out in debate. But though all of this is right, it is right only to the most limited degree because the performance of discursive difference precisely depends on the performance of discursive similarity. The making of difference, the kind of difference performed in decision making, thus demands and rests upon, the possibility of *accountable similarity*. That which cannot be said, or at any rate cannot be said in the right place, removes itself from the place of the Political becoming something quite Other.[12]

(2) This follows from the first point. If decision making tends to perform itself as the cockpit of important difference, then it performs not only a distribution between what may be said in important places where big decisions are made and what may not, but it also denies that anything "important" (or, for instance, politically serious) has been effaced. In other words, it performs most of the relations in the world as Unspeakable because they are "technical," a matter of "detail," or "aesthetic," or "personal," or because they belong to the realm of "fiction" or whatever.

And this is the second great technology of important decision making, another product of interference and collaborative overlap between different performances. It is one that we have come across in other guises—for instance, in the form of delegation of the pictorial into the "merely illustrative." But now we can see that it increases its size by effacing the fact that it effaces almost everything that might in another world be counted as important.[13] Or, to put it a little differently, in the context of big important Politics it deletes almost all of what we might call "the political" when this is understood as a texture of distributions and distributive possibilities performed in and through all relations.

The Politics of Decisions 3.
Collusions about the Importance of Place

Real decision makers are (made) powerful. For instance, they command obedience and, then again, they make "decisions." Both of these traits imply the performance of further distributions that have to do with agency. For real decision makers are made as agents. They act, and, in the extreme case, they are not acted upon. This may sound obvious but should not be taken for granted—for there are other contexts in which, for instance, the TSR2 is endowed with the power to act. We have seen circumstances and locations in which the technology, the machine, is performed as mobile, active, virile. We have seen how virility is built by distributing to passivity features of culture such as other agents (the enemy or the home) or parts of nature (such as landmarks or clearings in the woods), which (since the effect is one of contrast) means that these wait, wait to be acted upon by the aircraft.

But this is not what is happening here. The distribution is quite the other way round. It is *people,* specific people or particular *collectivities* of people, who are being performed as active. So we have cabinet ministers, these are made to be active. And then we have the cabinet itself, which is certainly being performed as an entity with the power to act. At the same time various aircraft, and in particular the TSR2 (but no doubt such other actors as the F111A) are being rendered passive as they wait for the decisions of cabinet ministers and the cabinet. Such, at any rate, is one of the distributions being performed in all the early exhibits. Indeed, so thoroughly and pervasively is it being performed (in this world who could *imagine* an aircraft making a decision?) that it is never said in as many words, but simply taken for granted.[14]

So to talk of "decisions" is to perform "decision makers"—here politicians—as agents. They act, but they also act in the right place at the right time; for otherwise their acts are ineffectual or they are not "important decisions." And here, to be sure, the right place at the right time means "the cabinet" or (to make similarities out of differences) "the government" (or "Her Majesty's Government"). To borrow a phrase from actor-network theory, these are the obligatory points of passage fashioned to be the center of the political universe, the places through which everything is made to pass.[15]

This is all very straightforward. Indeed it is obvious to the point of banality. The problem is that its very banality tends to deaden our critical faculties. So we need to remind ourselves that "the right place" and the "right time" are not given in the order of things, but that they are rather conditions of possibility made within sets of relations, generated in difference.[16] So, like the other differences we have discussed, powerful places are to be understood as the effects of the interferences between distributive performances, even (or perhaps one might add especially) in the divisions performed by those who do not like what they hear about the decisions emanating from those times and places.

Look, for instance, at the distribution performed in exhibit 7.8, which is a parliamentary motion from the Conservative opposition

---

EXHIBIT 7.8 "[I beg to move] that this house deplores the action of Her Majesty's Government in cancelling the TSR2 project." (Crossman 1975, 132)

---

party. It objects to the cancellation of the TSR2, which means that indeed it makes a difference. It makes a difference between government and opposition. But, at the same time, it performs "Her Majesty's Government" in a way that would excite no dissent from Her Majesty's Government's most partisan supporter. So "Her Majesty's Government" is being made as an obligatory point of passage, the relevant obligatory point of passage. In this way of telling, nowhere else is it right to perform the cancellation of the TSR2: the government is made, assumed, to be the place where it is *possible* to perform that cancellation, where it is appropriate so to do. And all this is being done in a performance made by the "loyal opposition."

This, then, is the third great interference or overlap that produces important decision making and thus the decisions of High Politics. It is a performance of place, of sociotechnical location. The effect is to produce a distribution between center and periphery, and to efface the possibility that there are other locations that might escape the gravitational pull of that center—or, indeed, the possibility that the world might perform itself without the need of special centers. But such a thought is, as they say, a fantasy.

The Politics of Decisions 4.
"Important Decisions" and "Mere Detail"

Now I encounter a methodological problem. This is because I want to make an argument about discretion. I want to argue that the big places where "decisions" are taken, make themselves, are made, discretionary.[17] As a part of this I also want to suggest that such places are performed as seeing further, that they are turned into places where matters are centered or (to use Bruno Latour's phrase) "drawn together" (Latour 1990). Or at any rate, I want to say that they are performed as having the capacity to act in a far-seeing discretionary manner, even if they are sometimes said to get it wrong, which is what the Conservative opposition was claiming about the decision to cancel the TSR2. To use a jargon, they are created as centers of translation or calculation.

Discretion and its performance: look again at exhibits 7.9 and 7.10 (the two form a pair). Crossman's complaint is that he is a cabinet

---

EXHIBIT 7.9 "The papers are full of reports about the TSR2 and discussion of whether we are going to cancel it or not. Day after day I read this in my morning paper but as a member of the Cabinet *I know absolutely nothing about it.* Even on Thursday when we had Cabinet the issue wasn't discussed. I read in the papers that it was being discussed in Chequers this weekend, with George Wigg and his pals present and people like me completely excluded." (Crossman 1975, 132)

EXHIBIT 7.10 "Actually, I am not against what is going on and I shan't complain; but it is true that when the issue comes up to Cabinet for final decision, those of us who are not departmentally concerned will be unable to form any opinion at all." (Crossman 1975, 132)

---

minister and that no doubt he will have to vote about the TSR2 one way or another, but that he is not where the action is (which is a version of the point made earlier about central places). In particular, he is complaining that he is not where the necessary information is to be found, which means that he is not being performed as a proper, discretionary, decision maker, someone who can take an overview and weigh up the merits of the options. He will not (he says somewhat inconsistently, having just expressed a quite specific view) be able to form "any opinion at all."

So Crossman describes something about the proper performance of discretion and its location, and performs himself in a different, nondiscretionary place. But once again these performances overlap, for there are endless examples allocating discretion to the cabinet, to government. For instance in exhibit 7.8, the censure of the opposition assumes that the government could have acted otherwise and retained the TSR2. So, though the difference between government and opposition is real enough, it rests upon the performance of a shared assumption: that the government indeed has discretion in this matter.

But what is the methodological problem? The *methods* by which cabinet ministers or cabinets are generated as discretionary centers of calculation is difficult to uncover and would require a study unlike the one I'm attempting here. The two points are somewhat related, but I will deal with the second first.

The problem may be succinctly stated. We are here located in the public domain and are watching the performance of a more or less public Politics. To be sure, the boundary between what is "public" and what is "confidential" is a construction (I explored a closely related distinction in chapter 2) and is, even when built, always blurred and subject to renegotiation (Crossman's diaries record and reproduce discussions that are scarcely "public"). Nevertheless, what I am *not* doing here is offering the ethnographic or historical material that would be needed to show how the space of discretion—the various aircraft options—is built. I'm not exploring how the discursive arguments that rank them are constructed as an effect of the distributions recursively performed within the networks of the administrative apparatus. Instead, I am simply reproducing a very small portion of that apparatus, which, let it be noted, in some measure reproduces the position in which the discretionary cabinet ministers find themselves as they wade through their briefs (see exhibit 7.11 which is by the Ministry of Defence Chief Scientific Advisor).

This, then, is the first version of the methodological problem. I am

EXHIBIT 7.11 "A few weeks after . . . [Denis Healey] took over, he asked me . . . for a personal appreciation of the TSR2 project, of which, when in opposition, he had been highly critical. As a basis, I used the report that I had prepared for Watkinson [an earlier Minister of Defence], amending it in accordance with what I had learned in the three years that had followed, and consulting only my own files. Healey went over the report line by line, with me at his side, and in my mind's eye I can still see him underlining passages." (Zuckerman 1988, 219)

neither in the right place nor undertaking the right kind of study if I want to tell stories about the administrative performance of the cabinet as a center of translation or explore the performative character of what is sometimes referred to as "governmentality."[18]

But there is a second and more interesting way of conceiving of the difficulty. This is to say that I am running aground on another distribution that is relevant to the performance of decision making—and in particular the decision making of High Politics. Indeed, the traces of such a difficulty are not hard to find in what I have been talking about. For instance, I mention "the public domain" and contrast this with what is "confidential"; and then I talk of the need for a "detailed" ethnographic or historical study; and finally I refer to the "administrative apparatus." But these are distributive tropes that come straight out of the discourses that perform a centered version of High Politics. And, like the other instances we have looked at, they perform their distributions asymmetrically in at least two different ways.

First, the division between the public and the confidential operates to (try to) conceal almost everything that might be said about the *basis* of government and, in particular, about the way in which High Political discretion is generated. True, as mentioned earlier, the boundary between the public and what is "properly" confidential is permeable. Crossman's diaries breach the divide (though they did not do so at the time the events were taking place). And, more generally, the talk of "leaks" bears witness to the frequency with which the divide is breached. But this very way of talking strengthens my point because it also *performs* the division between that which should be public and that which should not. For (as is obvious) a leak is matter out of place, a displacement of secret fluid that should have stayed in its container.[19]

Second, the division between political decision making and administration operates, in a hierarchical manner, to distinguish between that which is Politically important in terms of Big Decisions, and that which is not. "Operations," "administration," "accountancy," "technology," these are terms of contrast. They stand in contrast with "Political decision making," and this is a contrast that works in at least two ways. First, it works to efface the politically distributive character of technology, administration, and all the rest by implying that these are essentially nonpolitical. This reproduces another version of one of the distributions discussed earlier—the performance of a narrow and specific version of the political, one that indeed limits itself to High Politics. And second (which perhaps amounts to much the same thing) it relegates that which is not told as important decision making to a ghetto, a ghetto that is henceforth called "detail." We have encountered this before in several different forms, for instance, in the division between technics and aesthetics and in the organization of the brochure. But this time the divide is posed, at least in part, in terms of interest. It is posed in terms of what counts as interesting and what does not. This argument tells of and performs the commanding heights of Political decision making while relegating to the distant foothills of detail such routine matters as administration, technology, or illustration.

Discretion and discretionary places are created in a ramifying network of representational distributions. They do not exist in and of themselves. But the way in which such representational distributions perform discretion is in large measure concealed, performed as nonpolitical, and imagined as essentially uninteresting technical "detail." So this is the fourth distribution of decision making, another place of overlap and interference. It is the performance of a distinction between means and ends that graces important decisions, including those of High Politics, with a special place at the top of the greasy pole where the big and important decisions are taken, while effacing all the routines, the politics, that make this distribution possible in the first place.[20]

The Politics of Decisions 5. Effacing Difference

Earlier I made an assumption about the overlaps between difference narratives or performances. I assumed that the various "options" per-

TABLE 7.1

Decisions	Ex. 7.1	Ex. 7.2	Ex. 7.3	Ex. 7.4	Ex. 7.5
Cancel outright	✓	✓	✓		✓
Cancel and order F111A	✓			✓	
Cancel and take option on F111A		✓			✓
Cancel and order F111A and Phantom			✓		
Cancel and order a British aircraft	✓				✓
Make no decision until strategic review	✓	✓			
Continue with TSR2			✓	✓	

formed in the early exhibits may be mapped onto one another, that they are indeed sufficiently similar that their differences may be ignored. But now I want to ask whether this is right. And then I want to pose a much more interesting question that follows from this: If we ignore or, alternatively, attend to the differences between the narratives, then what exactly are we doing? This matter takes us to a theme that has recurred in a number of guises: to the politics of difference and their relationship to similarity.[21]

Table 7.1 performs a similarity by displaying the various options performed by the cabinet in the period March to April 1965—exhibits 7.1 to 7.5. I shall inquire into the basis of the similarity performed by the table shortly, into what is being done or effaced, in making this list. But first I want to look at the differences that it performs—differences that I earlier more or less elided.

Perhaps there are two ways of treating the divergence between that earlier listing and this table. One is as a question of method. We might argue that the shorter listing was flawed because it ran together important differences between options. Or, as against this, we might argue that the table is unnecessarily fussy. For instance we might insist that some of the distinctions that it draws rest on an unduly literal reading of the various exhibits. Thus we might say that Crossman's description (exhibit 7.1) of Healey's position on the F111A is really consistent with that of taking an *option* on the F111A—and if this is the case then we can collapse these two options together.

The fact is that there is no right answer: any possibility is defeasible in principle.[22] Nevertheless, the disagreement may be understood in two broad ways. On the one hand, we may imagine trying to create a better narrative, one that more closely accords with the events as these actually took place. In this case we treat it, so to speak, as an issue of method or epistemology, which is what I have been doing in the preceding paragraph. On the other, we may ask what would happen if we abandoned the idea that the exhibits describe a single set of options and instead stick with the idea that they are *performing* different distributions.

The first approach distinguishes between realities and representations, so it treats the various exhibits, in the way I described earlier, as *perspectives:* perspectives on a particular event or process, the cabinet meeting that examined the options, or the options themselves, distributions as they actually were. Historians work in this way daily, and so do detectives, journalists, sociologists, and students of technoscience. We all do so, for different perspectives are to be expected. But why would there be different perspectives? A number of responses suggest themselves—and are commonly deployed:

1. People may *forget* what happened—for instance, that it was an *option* on the F111A that was being sought by the government rather than an outright purchase.

2. They may perform differences as *unimportant* or irrelevant— again the difference between option and purchase. This would fit with a theory of social interests, one which says that knowledge is shaped by social concerns. Looked at in this way some matters, some differences, are simply uninteresting from a given standpoint.

3. They may *not know* fully what was going, being located, for instance, in the public domain rather than in the domain of confidentiality. Perhaps Healey's biographers are in this position.

4. They may deliberately *obscure* the facts. Exhibit 7.5, the press release, certainly does not say everything that it might have said about the background for cancellation. (Such an explanation would again be consistent with an account in terms of social interests.)

5. And finally, *circumstances change* so what appears to be contradiction may simply represent change. Indeed the difference between exhibit 7.9 on the one hand and exhibits 7.1 and 7.2 on the other may be understood in this way (exhibit 7.9 dates from January 17, 1965 and the others from April 1).

These moves explain difference by assuming that behind difference there is in fact a unity—for instance, in the form of a single cabinet meeting, a single set of options, a single distribution. They assume and perform the perspectivalism discussed earlier; that is, they assume that more or less adequate perspectives can be obtained on events and objects that are out there and independent of their descriptions. We are thus in the realm of epistemology and of method. We are in the business of assessing which description or combination of descriptions is most satisfactory and is most likely to accord with what really went on.

All of this is standard in the social sciences, not to mention life. But now we might note this: these perspectival, methodological, or epistemological moves would work equally well precisely to conceal *lack* of unity, to conceal the possibility that there is difference (as one might say) all the way down—and to efface the prospect that there is nothing out there that is independent of the methods through which it is described. Which, to be sure, makes the move toward ontology and performativity developed in chapter 2—while explaining why it is that matters appear to have to do with epistemology, perspective, and method.

In this way of thinking the world, the worlds, are being made in interference between performances and narratives. They are being made, in part, in coordinations or resonances between performances and narratives. And this is what we have witnessed here—for similar distributions make themselves through the various exhibits that I have discussed. Listing, generating discretion, distinguishing between reality and fantasy, effacing that which is turned into "detail," and then effacing the fact that there is effacing—all of these are performed in the more specific narratives and allocations of the cancellation decision. And the differences between these narratives, real though they are, tend to distract attention from their commonalities: from their tendency to enact similar ontological work—that of making a discretionary center.

But there is something more to be said. The hypothesis would be that places of discretion, decision-making centers, exist because, in the kind of oscillatory motion I have explored in earlier chapters, they are able to enact a distribution between the performance of narrative coherence and simplicity on the one hand and noncoherence and multiplicity on the other. This argument comes in two parts. First, such centers (appear to) make firm decisions because they (appear to) draw things together in a coherent manner—and indeed they do so, because that is what any *particular* performance enacts. Richard Crossman is clear enough about the options and so too is Harold Wilson, which suggests that firm decisions are indeed being made. It is just that their lists do not coincide. But this is the second point—*they are able to make firm decisions at all because they are, indeed, performing many different decisions.* All at the same time, and in parallel, decisions that are then coordinated and performed *as if they were the same,* as if they were a singularity.[23]

I am being cautious. I am not saying that what is being coordinated—all these different decisions or lists of options—is *in*coherent. To say so would be to make a move within the distributions of centered decision making. Rather I am saying that it is *non*coherent, that it is complex, and that part of the politics of centered decision making rests on this by now familiar double trick of managing the simultaneous performance of singularity and multiplicity, of, so to speak, *being* singular while *performing* multiplicity, or (it works equally well the other way round) of being multiple while performing singularity. So I am suggesting that here there is a kind of double play, a double looseness, another form of heterogeneity, another version of absence/presence, the simultaneous performance of solidity and fluidity.[24] At any rate, the possibility of "decision making" and, indeed, of High Politics rests in an interference, an overlap, where the performance of similarity depends on difference, and the performance of difference depends on the enactment of similarity.

Such, then, is one strategy of coordination or (it amounts to the same thing) one mode of interference. An interference that makes a fractional object, a decision, that is more than one but less than many.

End Words

In this chapter I have described some of the politics of decision making, and in particular of High Political decision making. In doing so I have set "politics" with a small *p* against "Politics" with a capital *P*. I have argued that big important decision making may be understood as a somewhat overlapping set of strategically and asymmetrically ordered performances that enacts a distribution about what is to be acted as political and what is not. It thus legislates most of what (after Michel Foucault) we think of as the *textures* or the *microphysics* of the political out of Politics. And it also effaces the fact that it is doing so, thereby rendering other possible versions of politics, other kinds of relations, fantastic, unpolitical, irrelevant, unimportant, or incoherent and hence, unperformable; they are not in the right place at the right time because they do not perform themselves within the great cockpits of debate and contest, those special places of disagreement made within organizations of all kinds, including Politics.

Can we escape the asymmetries performed by the bias to the center? Let's admit that this is difficult, for these are real effects, these asymmetries. They are real effects that perform themselves in many places and in many different and interfering modalities. They do so in words, but also in concrete, steel, titanium, in the actions of policemen and students of economics, sociology, politics, and technoscience. They do so in a range of different genres. So they are real enough, and they cannot be wished away. They have, instead, to be *performed* away. So I repeat the question. Can we escape the asymmetries of the distributions performed by the bias to the center? Might we perform them away? For if we were to do so, we would discover other political worlds to be thought and made, thought and lived.

I believe that the answer is yes, but with difficulty. In the places where noncoherence butts up against coherence, in those places where it can be turned against coherence, slowly the tools are being made, the tools that begin to erode the clean and simple asymmetries of the distribution to the center and detect and decode the erasures that generate centering. These tools will restore difference, multiplicity, and—most important and most difficult—the oscillations of fractionality. We can tell stories of precursors, in which case I would choose to tell the story of Michel Foucault who discovered or created the contemporary epistome. But we can also tell stories that are closer to home, for in technoscience studies we too are making forms of distribution that begin to escape the methods of centering, alternative ways of knowing that are also alternative forms of politics. And these politics or orderings come, as one might expect, in the form of narratives that only partly overlap, as distributions that (per)form only partial connections.

These forms of politics, these forms of ordering? They acknowledge rather than repress the noncoherence of multiplicity and difference — as in the work of Annemarie Mol. They perform monstrous and partially connected beings into new kinds of realities — as in the cyborgs and coyotes of Donna Haraway, the fractional and holographic persons of Marilyn Strathern, or the quasi-objects, neither human nor nonhuman, of Bruno Latour. They play in the places between fantasy and reality by translating the epistemic imaginaries of the Australian aborigines — as in the work Helen Verran. They exist in decentered indigenous knowledge traditions — as explored by David Turnbull. They oscillate through ambivalences and cohesions in the health initiatives explored by Vicky Singleton, Anni Dugdale, and Ingunn Moser. Or they dance with great effort — as in the body ontologies decried by Charis Cussins.[25]

So there are spaces, diverse places for performing distributed and interconnected relations. Relations that do not collude with the centers made by or for decision making in or outside High Politics. Alternative politics that put aside the tired questions of epistemology and begin to imagine worlds where knowing and being recognize the complexities of the ways in which they overlap and interfere, celebrate their performativity, and take responsibility for the fact that they are also ontological.

You set about opposing the rhizome to trees. And trees are not a metaphor at all, but an image of thought, a functioning, a whole apparatus that is planted in thought in order to make it go in a straight line and produce the famous correct ideas. There are all sorts of characteristics in the tree: there is a point of origin, seed, or centre; it is a binary machine or principle of dichotomy, with its perpetually divided and reproducing branchings, its points of arborescence; it is an axis of rotation which organizes things in a circle, and the circles round the centre; it is a structure, a system of points and positions which fix all of the possible [*sic*] within a grid, a hierarchical system or transmission of orders with a central instance and recapitulative memory; it has a future and a past, roots and a peak, a whole history, an evolution, a development; it can be cut up by cuts which are said to be significant in so far as they follow its arborescences, its branchings, its concentricities, its moments of development. Now, there is no doubt that trees are planted in our heads: the tree of life, the tree of knowledge, etc. The whole world demands roots. Power is always arborescent. — Claire Parnet in Gilles Deleuze and Claire Parnet, *Dialogues*

The narratives and enactments of decision making perform, and at the same time presuppose, conditions of possibility. They distinguish between — and demand the distinction between — reality and fantasy. They efface what, after Foucault, we have come to think of as the "microphysics" of power, while simultaneously presupposing its operation. They enact and presuppose that there are special and privileged Political places. They distribute between what is henceforth to be imagined as important and what is relegated to the supporting role of mere detail. And they presuppose — and indeed require — the singularity of decision making while effacing what they equally require for singularity, namely its simultaneous multiplicity.

Behind this, then, there are two related suggestions. The first is a version of the argument I have made throughout and concerns the coherence of the oscillation between singularity and multiplicity and the interferences that it entails. This, then, is the trick of modern/postmodern alternation and slippage. But the second has to do with what one might think of as the "collusive" character of the interferences between multiplicities: how they efface the ontological work that they perform, and how they conceal the way in which they reenact the conditions of singular possibility. "Collusion" is a strong word, and I need to be clear that I am not accusing those who tell stories of bad faith. Instead I am interested in the ways narrative framings enact and reenact themselves—and this is the issue that I attend to, in particular, in this chapter. I argue that (apparently) singular narratives collude to produce a (seemingly) singular world with certain attributes such as chronology and scale, a world populated by (ostensibly) singular sets of objects, and that these conditions of possibility are made rather than given in the order of things. As a part of this argument, I explore the performative character of both academic and nonacademic storytelling more systematically and use the distinction between arborescences (which are grand narratives), and rhizomes (which look more like a tissue of little narratives). First, then, a grand narrative.[1]

A Grand Narrative

There's a section of the RAF called the Operational Requirements Branch. We've come across it from time to time along the way. It isn't very large. Or at any rate, in 1954, it wasn't very large. The job of the officers seconded to the Operational Requirements Branch was to think long term, to think about the needs of the RAF ten or more years ahead, and to try to imagine the form that "the threat" might take geopolitically—which as we know at that time meant the Russians and their allies in the Warsaw Pact, together with the possible aspirations of new powers in what is now called the "Third World." They also thought technologically—in terms, that is, of the likely innovations that would be made, in particular by the most advanced of these threats, the Soviet Union.

There was talk throughout the RAF. But papers defining a change in the threat started to emerge from the OR Branch between 1953 and 1955, and what they said is this: the life of the British nuclear bomber force is limited.

At this point, to make the narrative work perhaps I need to offer some background, to make some context, some scenery.

Britain was in the process of becoming a nuclear power. It was working toward atomic weapons and was also planning hydrogen weapons. These were free-fall bombs, weapons designed, in the first instance, to play a strategic role, though they would later be developed for tactical use, to be carried by Canberras. The idea was deterrence. An enemy could be held back by threatening to drop bombs on major cities. These bombs were to be carried by aircraft, by the so-called V-bomber force. We have come across the V-bombers. These were subsonic, medium-altitude aircraft that would fly from bases in the UK or Germany to the Soviet Union. Their existence, together with the nuclear weapons themselves, provided the British nuclear deterrent.

The life of the British nuclear bomber force was limited. Why? This is something else I have discussed. The OR Branch feared the advent of surface-to-air, radar-guided missiles. These would, or so it imagined it, be located around major targets such as Moscow. And as the technology developed, Britain's V-bomber forces would become increasingly vulnerable. The Soviet Union was developing surface-to-air missiles, but the technology would take some time to mature and even longer before such missiles were deployed in numbers sufficiently large to make any substantial difference. Nevertheless, the writing was on the wall. By the mid 1960s, ten or more years off, the V-bombers would have become obsolete and Britain would no longer have a nuclear deterrent—unless something was done.

This, then, is one of the possible branches. Perhaps it isn't the most immediate for the aircraft project that was to become the TSR2—the aircraft project first called General Operational Requirement (GOR) 339 and subsequently by the more specific Operational Requirement (OR) 343. But it's important even so.

Unless something was done—but what? If the UK wanted to retain its nuclear deterrent (and this was never in doubt, at least in the Royal Air Force and government) then perhaps there were three possibilities.

First, it might develop ballistic missiles. Launched from the UK or Germany, or in some versions from aircraft based in these countries, such missiles would be targeted on the great cities of the Soviet Union. And, because of their speed, like the Nazi V2 missiles they would be quite immune to counterattack by surface-to-air missiles.

Second, it might develop a very high-speed, manned bomber. This aircraft, which would also fly at high altitude, would be too fast for surface-to-air counterattack. And, like the first option, would (or so it was suggested) outfly any likely fighter aircraft that were sent up to intercept it.

Third, it might develop an aircraft that would fly very low. So what was the rationale? Such an aircraft would be difficult to spot. If the surface-to-air missiles, or for that matter the interceptor fighters, were being guided by radar then the radar needed to be able to "see" the incoming attack aircraft. But radar could only see long distances at medium or high altitudes. Because it worked only along direct lines of sight, and the curvature of the earth, together with unevenness in the terrain, meant that an aircraft flying very low would not be detected until the last moment.

Each of these possibilities was to be explored. Each was, to some extent, to be developed. But the project that embodied the second option was canceled at a very early stage, which left the first and the third—and the third was to turn into GOR 339.

With some modifications. For instance, there isn't any need to fly low all the way from Germany to the Soviet Union. A radar around Moscow can't see into Germany and, in any case, it also uses a lot of fuel to fly at two hundred feet. So the proposal became this: the aircraft should cruise subsonically and economically at medium altitude while far from its target; then it should fly very high and very fast (Mach 2), before descending to low-altitude subsonic or transonic flight for the approach to the target.

How much detail do we need? How long is a story? How long is a piece of string? How much context is necessary, organizational, strategic, or technical? I've told something of each of these. Now here's another branch to the story. It has to do with "procurement." As I noted in an earlier chapter, procurement is the word they use to talk about procedures for conceiving, designing, and in particular acquiring military aircraft.

In 1945 Britain was still a great power. It had been exhausted by years of war, but it still had a global role, global colonial possessions, and global military responsibilities and ambitions. Only with benefit of hindsight was the established wisdom to become that the world was changing, that the economic dominance necessary to sustain Empire was no longer a reality, and that geopolitics were shifting in favor of the Soviet Union and especially the United States.

So there was a combination of factors: there was the need for global military reach; the years of a war economy that gave the highest priority to procurement; the relative cheapness of Second World War aviation technology, at least compared with what was to follow; the existence of many aircraft manufacturers in the UK, again built up and sustained by urgent need in the Second World War. This combination led to a particular pattern of aircraft procurement. The government would order prototypes from several companies expecting that they would be tested and improved. Then, the most satisfactory—or in some cases several of the most satisfactory—would be put into production. This is one of the reasons that there were three quite different types of V-bombers, not just one.

In 1945 the government guessed that a major military threat was at least a decade off, which meant that there was plenty of time to undertake major projects, slowly. They called this the "long step" approach: big technological advances and long time horizons. But then, in 1951 came the Korean War and with it the fear of another global war. This concern led temporarily to a large increase in the numbers of aircraft ordered. Prototypes were rushed into production with sometimes troublesome results. For instance, large numbers of at least one type, the Swift, were never properly put into service at all because the technical problems were not resolved. It was also discovered that although aircraft flew satisfactorily without equipment, they encountered difficulties when armaments were installed or fired. The Hawker Hunter was a case in point.

What was the diagnosis? What was the character of the procurement problem—apart, that is, from the rush to rearm brought on by the Korean War? The answer to these questions led in 1955 to two policy decisions about the procurement of military aircraft. We have encountered the first of these already: that aircraft should be treated as "weapons systems." They should be designed and built as an integrated system, including weapons, equipment, and airframe, rather than as a "weapons platform" for carrying weapons that were to be bolted on, as it were, afterward —at which point experience suggested that insuperable difficulties might emerge. The second was that the development and the production processes should be integrated. A large "development batch" of aircraft would allow problems to be identified and resolved more quickly, thereby eliminating the bottleneck caused by a limited number of prototypes, and the development aircraft could then be brought up to standard and introduced into service.

Stories about procurement methods. Now back to strategy.

In 1957 the Minister of Defence was Duncan Sandys, a man we have already encountered. Sandys wasn't much admired by most people in the RAF. They talked, behind his back, of the "shifting, whispering Sandys," and we have already seen the reason for this—it was that he was committed to missiles. These, he believed, were the technology of the future. Because the United States and the Soviet Union were developing them, it was appropriate that Britain should do so too. So he issued a government policy statement, a "White Paper," in which he boldly announced that the government would no longer develop most forms of military aircraft. There was certainly no room for manned bombers because the UK would henceforth commit itself to ballistic missiles in order to secure its nuclear deterrent.

And what of GOR 339?

Now I need to provide some more context. I need to go into the question of tactical warfare, into a line of storytelling that will displace the arguments about nuclear deterrence and procurement. I
need to describe how GOR 339 really had more to do with tacti-

cal strike and reconnaissance aircraft than with the grand strategic themes I've mentioned so far. Tactical strike and reconnaissance: hence the acronym TSR.

Tactical support. This meant bombing bridges, roads, railways, factories, ports, columns of tanks, and airfields. Now we need to distinguish between two forms of tactical support, one on or over the battlefield and the other in the form (to use the jargon) of "deep interdiction," which involved bombing targets such as those I've just listed, which are miles, even hundreds of miles, behind the battlefield. We have seen that there was an aircraft in service that played this role. This was the Canberra. But it couldn't fly terribly low and it was subsonic, which meant that many of the same arguments about the vulnerability of nuclear bombers also applied to it. Hence there was need for a "Canberra replacement," the GOR 339 aircraft, to play the role of deep interdiction.

Another branch in the story is reconnaissance. There was need, or so they said in the OR Branch, for an aircraft that could reconnoiter, look and see, and detect military movements and concentrations, again, miles behind the front line. In an ideal world with limitless resources this would be another project, a different aircraft altogether. But how about combining this requirement with deep interdiction? This was the OR Branch proposal.

Now there is yet another branch in the story, one that I earlier left in the air. This has to do with what we now call the Third World.

Much of the Third World was, at the time, part of the British Empire or its client states. We have come across this role, this need to fly, this responsibility for acting "East of Suez," in countries such as Malaya or Aden. But there wasn't any need for nuclear deterrence in the "colonial theater." Third World countries did not yet have nuclear weapons. There was plenty of room, however, for aircraft that could attack tactical targets and undertake reconnaissance in so-called "brush fire" wars. This, at any rate, was the argument of the OR Branch as it reflected on the lengthening shelf life of the aircraft that were currently playing that role.

Now we can rejoin the main narrative. Deterrence and ballistic missiles, tactical strike, reconnaissance, procurement and organization: as we have seen, the stories start to come together.

April 1957. This is a crisis point for the RAF in general and for GOR 339 in particular. The Minister of Defence, Duncan Sandys, has announced the cancellation of all new bombers and fighters for the RAF. They will be replaced by ballistic missiles. One or two aircraft types have crept through. One, designed to fly from aircraft carriers, a machine we met in chapter 4 and will encounter again, is called the NA 39 or the Buccaneer. But what is the future for GOR 339? This is an aircraft that at this stage isn't even properly on, let alone off, the drawing board.

If you look at the official papers for 1957, they reveal a flurry of correspondence between, for instance, the Minister for Air (who was in charge of the RAF) and the Minister of Defence (who was responsible for all the armed forces). When he promulgated his policy about missiles, the Minister of Defence thought he was canceling all RAF combat aircraft. The Minister for Air thought that the reconnaissance GOR 339 was to be saved—though as with the cancellation decision discussed in the previous chapter, the fact that they were not taking the same decision was not clear at the time. After a flurry of official paperwork and a good deal of sweat, the Minister of Defence ended up by saying, I can be reluctantly persuaded that missiles can't act in a reconnaissance and tactical strike role. But why do we need a *new* type of aircraft for this role? Why do we need a new type of aircraft when the Royal Navy already has its Buccaneer under development and indeed, this is at quite an advanced stage? Would this not do for the air force as well?

So this is another actor, another branch in the story. That of the Buccaneer. But how much do we need to know? In the present context, perhaps not so much. Let's just say that what followed, the negotiations that took place in the corridors of the government machine (negotiations which included aircraft manufacturers and the procurement branch of the government, confusingly called the Ministry of Supply and later the Ministry of Aviation) fits the kind of pattern imagined by those who study bureaucratic politics. Moves that catch the flavor of the 1957 debates would look like this.

Navy: You ought to use our aircraft, the Buccaneer.

Air Force: No. It's too small, vulnerable, and it's underpowered. It can't fly far enough. And it doesn't have the precision electronics we need.

Treasury: But it would be much cheaper to have one aircraft rather than two.

Ministry of Defence: Hear, hear!

Air Force (to Navy): All right. That makes sense. So how about *both* of us having GOR 339?

Navy: No way: Your aircraft is years off. Ours is already being made. In any case, yours is far too big to fit in an aircraft carrier.

Air Force: If you say so. But there is no way your small, slow aircraft will ever meet our specification. Two engines are essential for safety and reliability. We must have our own aircraft.

Vickers Armstrong: But we can make a small version of GOR 339 that would be just as good as the big one, and it would fit in aircraft carriers too!

Air Force: Certainly not! We've already said that our plane needs two engines!

Navy: Let's be serious. Your small GOR 339 won't see service for goodness knows how long while our Buccaneer is nearly ready. A plane in the hand is worth two . . .

Ministry of Aviation: Excuse me, but there's something else going on here. While you're bickering about the aircraft you want, we also need to be thinking about the future of the aircraft industry. It needs pruning, sure, but we shouldn't go too far . . . we need *some* orders.

(Short silence)

Ministry of Defence: All right. I don't like this disagreement between the Navy and the Air Force. Not at all. One aircraft would be much better than two. But if you absolutely insist you need a large aircraft for the Air Force then I suppose I've got no choice. I'm very unhappy about this, but reluctantly I'll let you build your GOR 339.

Ministry of Aviation: And it will be built by a consortium. This will allow us to rationalize the aircraft industry.

Treasury (aside): But we'll do our best to stop it along the way if we can, by putting up obstacles.

Parts of the Navy: And we'll do everything we can to put a spoke in the wheel too. The whole idea of another aircraft is a nonsense when we've got the Buccaneer.[2]

So in the autumn of 1957 the aircraft manufacturers were told that they should prepare outline designs for GOR 339. But they had to do so in collaboration with one another. Individual proposals were not allowed.

In January 1958 these designs were submitted. Toward the end of 1958, after close study and a great deal of further bureaucratic infighting, some of which I discussed in chapter 4, it was announced that a contract to design and manufacture a GOR 339 weapons system would be awarded to a consortium of two companies, English Electric and Vickers Armstrong. Subsequently, the relevant parts of these two companies were to merge to form the British Aircraft Corporation (later British Aerospace). It was indicated that Vickers would be the dominant partner in the GOR 339 project, both because it had more systems expertise and because it was believed that it had stronger management. And the expectation was that, at a subsequent point, a development batch would be ordered.

That is a narrative of the conception and design of the TSR2, some of the various branches that led to the decision to build. Subsequently the project, as they say, "progressed"—and a narrative of that progress might run as follows:

The doctrine of missile dependence did not long survive Duncan Sandys's quite brief incumbency at the Ministry of Defence. Britain abandoned its own attempt to build long-range ballistic missiles and after various vicissitudes, bought American Polaris missiles to carry British nuclear warheads and install in submarines built in the UK. At the same time the TSR2 began to play an increasingly strategic role in the minds of defense planners. In part, this was because the distinction between tactical and strategic nuclear weapons was becoming increasingly fuzzy in the 1960s as smaller nuclear weapons were becoming available.

The first TSR2 aircraft flew in September 1964, well over a year behind schedule and very substantially over budget. In the meanwhile it had become clear that no foreign government would buy the aircraft, and we know, of course, that in April 1965, after a change of government, the aircraft was canceled. An option was taken on an American aircraft with some similar properties, the General Dynamics F111A. This option was abandoned in 1967 as a result of a sterling currency crisis. In the end the RAF obtained a number of kinds of aircraft, none of which fully matched the specifications of the TSR2. These included a developed version of that original naval aircraft, the Buccaneer, which saw active service in the 1991 Gulf War, nearly forty years after it was first conceived.

Arborescence

This is the story that I *haven't* told you. It is the story that I have avoided as I have taken empirical cuts through the project in earlier chapters. On the other hand, it is the kind of story that I might have told had I wanted to write a "plain history," a social history, or a version of the social shaping of the TSR2 project. Though it catches something to call this story a grand narrative, this would no doubt be unfair, both to the story itself and to Jean-François Lyotard (1984b). So perhaps it would be better to find a way of pinning down something about its specificity. To do that we might use the language of Gilles Deleuze and Félix Guattari and call it an arborescence. In other words, it is a form of storytelling that is treelike in structure. "The first type of book," they write, "is the root-book. The tree is already the image of the world, or the root the image of the world-tree. This is the classical book, as noble, signifying, and subjective organic interiority (the strata of the book). The book imitates the world, as art imitates nature" (Deleuze and Guattari 1988, 5).

Deleuze and Guattari are playing on more registers than I want to handle here. But if the TSR2 story isn't a book—it isn't long enough for a start—then it is certainly arborescent. It is a root narrative that seeks to imitate the world. For as the story strings itself out across the lines and pages, it builds "a whole apparatus that is planted in thought in order to make it go in a straight line" and "a hierarchical system or transmission of orders." Things, events, and considerations are made to stand in relation to one another, asymmetrically, within a structure of branching points. Events govern other events. More distant branches come together to form a story and make a conclusion in the shape of, that shapes, the aircraft. Organization, strategy, technology, procurement, geopolitics, tactics, colonialism, bureaucratic politics—on this count there are at least eight major branches leading to "the decision" to build the TSR2 and then to build it in the particular shape that it had.

So like the tree of Jesse, the tree is hierarchical: one set of things, events, factors, is related to another. Layer is laid upon layer. But there is something else. This technoscience arborescence also reflects and maps the passage of time: "it has a future, a past, roots and a peak, a whole history, an evolution, a development." To say this is to note that it works by describing (some version of) cause and effect; that it tells what precedes what. True, there is also space. Simultaneity, different events, different branches or roots of the tree, the processes that make these up occur at the same time. The diachronic and the synchronic, they are both assumed in the story. So the arborescent narrative grows in, presupposes and creates, the kind of three- or four-dimensional Euclidean time/space container imagined in chapter 4. Like a bonsai tree, it's a smaller version, it tells a smaller version, of what there is already, out there. It grows continuities and coordinations in the form of roots, branches, and relations, but it also performs what these presuppose, the conditions of its possibility: on the one hand, the passage of time, the greatest hierarchy, the greatest asymmetry of them all; and, on the other, the machinations that spread themselves across space.

Performativity

How does telling stories make a difference? Having made an arborescent narrative that is also a description of the TSR2 project, this is the next question I want to explore.

In one way we already know the answer. It is built into the semiotics that subjects perform objects and objects perform subjects. It is built into the structure of the book. Telling always makes a difference of one kind or another. But if some stories make more of a difference

Cultural Bias

Reading an earlier version of this book, Bruno Latour suggested the following thought experiment.

Would it be possible, he wondered, to build a sense of "the project as a whole" out of a set of bits and pieces, little stories that had to do with this or that project-related matter? Would it be possible to build the present book in this way by offering a series of small stories and then, at the end, to have generated a chronological time-space story of "the project as a whole"?[3]

Until this last story I've largely resisted the arborescence of project narrative. So a possible question is whether you, the reader, have built a chronologically and spatially ordered narrative out of the bits and narratives that have composed the earlier chapters: whether you have supplied your own project-relevant arborescence.

If you have indeed done so then we have learned a little about the distributions of narrative, about the relations between what Deleuze and Guattari call the arborescent and the rhizomatic. Or, to put it in another language, we have learned something about the character of cultural bias: about a tendency to find smoothness and coherence, and the possible character that coherence might take. About our aversion to noncontinuity. Or noncoherence. Or to sensing the oscillations between continuity and discontinuity, the interferences that make singularity out of multiplicity in a movement that is continually deferred. Perhaps, then, we have learned something about the current performance of the conditions of possibility.

than others, then we need to think harder about interferences and the conditions of possibility. And if we start to ask questions of this kind—and in particular about interferences—then we blunder into a place inhabited by linguistic philosophers, and in particular a place that is occupied by the words "I do."

Philosopher J. L. Austin (1966; 1970, 235) argued that if these particular words are uttered at the right moment and in the right place, under what he called "felicitous" circumstances, then they are also actions and not just words. But what is the "right place"? In this context, it is a properly constituted marriage ceremony where, for instance, neither of the prospective partners is already married. Under these circumstances if I say "I do" at just the "right moment," I end up married. Such is his definition of the performative: it is a word or a set of words that is also an action.

Let's displace the example a little. Staying with the humanist theme of romance, if I say "I love you" to a person in "real life" (to be sure, the definition of what should count as real life is precisely one of the issues at stake), then this has another kind of effect. In Austin's terms it would be a constative; that is, it states something. It is a description, which means, at least in the paradigm case, that it is either true or false. But it is also in some sense performative, and this is what's important here. The question is, what is it performing? If it were said in good faith and all the rest, it would be a performance of love. But it might also (and here the uncertainties crowd in) be the start of a love affair, or (no doubt equally uncertain) its reaffirmation. Or, if it were said in the "wrong way" or under the "wrong circumstances," it might be the end of a love affair or a friendship.

So what do we learn? The answer is that to say "I love you" is probably performative too. Like "I do" it also makes a difference. It enacts something and it has force. But at the same time it is also clear that the focus of attention has started to shift from a relatively clear set of conditions that secure a specific form of performativity ("I do"), a specific outcome, to a whole lot of uncertainties both in degree and in quality—which uncertainties are interesting, indeed crucial, if we are concerned to make a difference and think about the *kind* of differences that are made when stories overlap.

Let's displace the example once more, or better, replace it with
something entirely different. If I say, "The government has fallen,"

then this is different again, isn't it? If I've just heard it from the BBC news (as opposed to a novel by Anthony Trollope), then something else is happening. Commonsensically, we could say that I'm reporting on a state of affairs or the affairs of state. This means that it is a constative and not a performative at all. Thus we could also add that what I say doesn't make much difference, indeed perhaps no difference at all, not, at any rate, to the government. This is because the toppling of the government has happened, as it were, "out there," and my words simply report on something. They are not (how to say this?) a *part* of the action. They perform, instead, a kind of perspectivalism and belong to epistemology instead of ontology.

Let's make one further displacement, again on the political theme to do with the falling of governments. If I say "the government will fall tomorrow," then here we have something different yet again. Once more it sounds a little like a report, a perspective taken on something that is out there—or will be out there. This sounds like the world of the constative, but the difference between "*is* out there" and "*will be* out there" is crucial, for now we are starting to move back toward the earlier declarations of love. The prospects are uncertain, and, as a part of that, the declaration itself may turn out to be performative—as it was when, for instance, tens of thousands of brave people stood and clinked their keys in Wenceslas Square in Prague on what turned out to be the eve of the Velvet Revolution. "Your time is up," they clinked. And the clinking (it turned out) performed the departure of that sad and vicious government. "The government will fall tomorrow": these words have become, albeit uncertainly, a part of the action. Constative and performative both, they are constative precisely *because* they are performative.

I don't much care for such armchair examples. Michel Foucault's work teaches us that philosophy is better pursued by empirical means, which is why I have written this book in the way that I have, as a book "about" an aircraft. But before we move on, look at what happens if we put the four "philosophers' examples" that I've created side by side (table 8.1).

This helps to make the story that I'm trying to tell clear, or so I hope. Sometimes words, stories, and no doubt pictures are *also* actions. That is, they make the worlds that they describe. And sometimes they aren't, and they don't. And then again (a somewhat differ-

TABLE 8.1 Performativity

	"I do"	"I love you"	"The government will fall tomorrow"	"The government has fallen"
Action (performative)	✓	✓	✓	
Report (constative)		✓	✓	✓

ent but equally important distinction), sometimes words and stories act in clear and unambiguous ways, and sometimes they don't. Such is the space that I would like us to investigate, the performativity of narrative, as overlap and interference with other narratives.

Back to the empirical.

Little Narratives

Perhaps we could agree, for the moment, that the arborescent project narrative isn't an action and say that it's simply a report.

Now look at this:

Sir Cyril, emphasising that he spoke with the full authority of the absent Minister, got straight to the point which was that only one military aircraft project had (so far) survived the 1957 White Paper, and that was OR 339—the Canberra replacement. There was no certainty of further military aircraft projects and, assuming OR 339 went ahead, the contract would only be placed with a group of firms, or with two or three firms acting in co-ordination and with one designated as the leader. The Government hoped for eventual rationalisation and amalgamations. . . . How the industry so grouped itself was its affair, but group itself it must.

The reaction of the heads of industry to this ultimatum was strong and virtually unanimous. For the industry to remain viable at all in any form, there had to be an on-going military programme, not just the rather dubious carrot of a possible Canberra replacement. Further, the airframe industry could not (repeat) *not* possibly exist on civil work alone. Sir Cyril said that there was certainly no hope on the military side unless there were major amalgamations or rationalisation on the lines he had sketched out as a condition for submissions for OR 339. He could only repeat that it was, therefore, up to the industry to arrange their own affairs and pick their own partners if they wished to put in a submission. (Gardner 1981, 23–24)

So here we have another aircraft story, a smaller narrative. It is a narrative that interacts with the arborescent story with which I started this chapter. It describes, as they say, "in more detail," the moment when the government told the aircraft industry that it should rationalize.[4] One of the branching points. So it's another report, another constative.

But shift focus. Attend instead to the story told to the industrialists by the big-shot civil servant, Cyril Musgrave, for this too is a story, a narrative. "It will," he was saying, "be like this. My story, the Musgrave story, is that this is how it will unfold, the future history of the British aircraft industry." These are constative words, but they are words that are also actions. They are words that will make and distribute, indeed redistribute social, economic, and industrial relations.

So there is a difference between my story and Gardner's on the one hand, and Musgrave's on the other. To say it briefly, Musgrave's story makes a difference, it's performative, whereas mine doesn't and isn't, and neither does (or is) Gardner's. They don't (as is sometimes said) "belong to the real world" but rather seek to describe it (though what we mean by the "real world," what counts as "reality," is precisely what is at stake).[5]

Well, perhaps Musgrave's story makes a difference, but only perhaps. Because the passage continues so: "After the meeting had ended, Denis Haviland, who was present as Under Secretary (Air), Ministry of Supply, recalls that Sir Cyril Musgrave turned to him and said two words, 'We've won'" (Gardner 1981, 24).

Interesting, isn't it? Musgrave watched the big-shot industrialists leave and then turned to his sidekick and said "We've won." This tells us, as is obvious, that he wasn't sure he was going to win and that things might have been different, that Musgrave's story might not have been performative in the way he intended. His story might not have redistributed the world, it might have had no effect or indeed had some other unintended consequence. We're discovering that at the time he gave it, his little speech didn't necessarily fall into the "I do" category but rather into the class of "I love yous," into the class of the "I love yous" that aspire to be turned into "I dos." Or the class of "the government will fall tomorrows" that hope to turn themselves into "the government has fallens." This is an aspiration that, as history (or the narratives that make history) tells us, was achieved by Musgrave. But this wasn't clear, not beforehand, not at the time.[6]

Making a Difference

Musgrave's story made a difference. It helped to bring what it described into being. If we want to pick through the entrails and ask how it achieved this, we could use various theoretical vocabularies for talking of interference effects. We might talk of intertextuality— for here, certainly, we have one form of talking that interacted with others, those, for instance, of the industrialists, to produce the effect of reordering the aircraft industry. Again, we might talk, like the actor-network theorists or Donna Haraway, of the performance of materio-semiotic relations,[7] and then we might remind ourselves that narratives work themselves through a range of materials. So we'd note that it certainly wasn't the honeyed power of Musgrave's voice that performed the industrialists into submission, that made the difference by reordering the distributions of British industry. He might just as well have croaked like a frog, and the outcome would still, might still, have been the same.

So be it. Such is the character of interference: it comes in many forms and operates in many ways. It may be told of too, performed, in many ways. But what about my arborescent narrative? What about the stories of the technoscience student? Or those of Gardner, the company historian? How are they working? And how do they intersect?

We agreed pro tem to lodge these narratives in the "report" pigeon hole, the place where we put "the government has fallen." This place is inhabited by constatives in the realm of epistemology. The reason is that, notwithstanding all the strictures about subjects and objects that make themselves together, these stories don't seem to be making any difference. They aren't acting upon, or better, within whatever they report.

But this distinction doesn't really work, does it?

"Was O.R.343 a Valid Concept?"

"Should a Short-Step Aircraft have been Produced?"

"Was a Sophisticated TSR Aircraft Necessary?"

"Was the Cancellation Justified?"

"Was a TSR-2 Aircraft Necessary in 1965?"

"Was the Aircraft Industry too Large?"

"Had TSR-2 to be Cancelled to Further European Airspace Collaboration?"

These are chapter and section headings taken from a small book called *Crisis in Procurement: A Case Study of the TSR-2*, which I've already cited several times (Williams, Gregory, and Simpson 1969). This publication appeared in 1969, four years after the cancellation, but the headings give the game away. They tell us that it is not simply a report. It is rather a document that's trying to answer policy questions.

Some more citations from the same source:

By quantifying the quantifiable, therefore, the 1965 TSR-2 decision can be shown to be correct, though the gains from its cancellation were so small that a slight alteration of estimates could easily have produced a contrary argument. However, when the unquantifiable elements are added, the issue becomes an even more open one, so much so that it is difficult to arrive at a clearcut answer by rational analysis. . . .

However, had the normal accounting conventions been altered in the way that has been suggested, and greater account taken of the increased reliability and availability of TSR-2 as a weapons system, it would have been valid to argue that the TSR-2 ought not to have been cancelled for the F111K. It is interesting to speculate, however, how far these detailed arguments were presented to the decision-makers concerned in terms of a dispassionate analysis of the alternative assumptions involved, and the related probable outcomes. If they were not, no truly informed decision was possible, and the decision taken would automatically be a bad one in decision-making terms. (Williams, Gregory, and Simpson 1969, 66)

So this is another recommendation about the distributions proper to centered decision making. Which is another way of saying that it is also a policy report: "In the preceding chapters the difficulties which beset the TSR-2 project have been analysed. . . . An attempt will now be made . . . to relate these difficulties to the production of future British weapons systems" (Williams, Gregory, and Simpson 1969, 67). It is a policy report that, no doubt, aims to make a difference. It is an "I do," or at the very least a version of "I love you" or "the government will fall." So it is all about performativity. It is about *making*. Like Musgrave, it's about uncertain performativity, a description of how it was, how it is, how it should be, how it might be. It's about the performative uncertainties that arise when (as is almost always the case) the character of the interferences between different stories is uncertain.

Williams and his friends are thus different from Musgrave. We want to say, don't we, that Musgrave was "more deeply involved," more of a participant than Williams and Co.? On the other hand, we're also learning, or so it appears, that overlap and participation is a gradation. Or better, we're learning that whether or not the author of a narrative—or perhaps better a narrative itself—participates in what is being narrated depends on how the line is drawn between inside and outside, which is perhaps a way of talking about how the overlaps between performances are built and rebuilt.[8] We might want to say, for instance, that Williams and his friends were outsiders to the TSR2 project, but not to the "defense policy community"—except, of course, we'd also need to add that such a distinction is itself performed, an effect. It is a patterning of narrative distributions that makes similarity and difference in the slippery place between "I do" and "I love you," between "the government will . . ." and "the government has . . . ," between the descriptions of simple epistemology and the world making of ontology. Which is, no doubt, where we all are, where all stories are to be found, multiply distributed in the fractional interferences between telling and doing.

So there are four narratives in play: the arborescent story of the social shaping of technology with which I opened this chapter; Charles Gardner's narrative, the story of the British Aircraft Corporation as written by the historian; Cyril Musgrave's story, the account offered by, or on behalf of, the senior civil servant; and there are the recommendations of the policy analysts, Williams and his friends. At which point the density of these intersections begins to become somewhat overwhelming. But that is how it is.

Intersections

The following citations come respectively from my summary story, Gardner's story about Musgrave, and Williams's account written with his friends.

> So he issued a government policy statement, a "White Paper," in which he boldly announced that the government would no longer develop most forms of military aircraft.

> The Government hoped for eventual rationalisation and amalgamations. . . . How the industry so grouped itself was its affair, but group itself it must.

> By quantifying the quantifiable, therefore, the 1965 TSR-2 decision can be shown to be correct.

We've established some of the differences between these. "Plain history," participation, policy, certainly they perform differences, and if they are performative then they perform different TSR2 projects. Multiplicities. But then the question we've been wrestling with all along recurs: Do they not also interact together in ways that tend to create a single object, a single project?

The question is rhetorical. The answer I want is that they indeed intersect. They intersect in terms of specificities. They coordinate themselves by talking of "the same" events, or "the same" project, but as they frame and perform these specificities they also make "the same" conditions of ontological possibility—singular conditions. I said that my narrative is arborescent in character, that it is like a bonsai tree, existing in and making a literal and metaphorical Euclidean space-time box. But if it is a little version of a large tree, an arborescence, then it also makes roots and branches. It performs places where different lines come together, junctions that are more or less important, closer to or more distant from the trunk. It makes bifurcations and confluences to create this developing project-plant.

In the way that I have told the story, I have attended most of all to the place where the decision was made to build the aircraft and to build it in a particular form. To a smaller extent I have also described the decision to cancel. The tree makes, most of all, a project beginning and a project end. The story that I told organized itself around these: the joining of the twigs into branches achieved their status, their salience, their hierarchical import, in relation to the beginning and the end.

The smaller narratives, however, are also built to the same pattern. It is possible to be more precise. They are all about the decision points that appeared in my narrative, important bifurcations. They all perform "aspects" of those decision points. They tell us, "in more detail" about places where particular twigs came together to form branches, or particular branches came together to form the trunk. They explain why the decisions took the form that they did, why the tree grew in this way rather than that. So no doubt it is possible to read them in other ways, but within the project-relevant, bonsai, space-time box of my arborescent narrative, these smaller narratives *perform the same pattern*. It is just that they do so at a higher level of magnification. Increase the magnification and what do we get? The same bifurcations at greater length. Or an expanding pattern of further bifurcations and sub-bifurcations that go toward making the larger branching point. Branchings and more branchings—the interferences between the different narratives are treelike in form. They make a stable pattern of interference as they overlap because they are self-similar. Branches go on appearing as magnification increases: in their structure they are scale independent.[9]

So time and space are arborescent effects. But so, too, is scale. Indeed, *space, time, and scale are made together*. To make a space-time box is to imagine the possibility of looking at the contents of the box "as a whole." As singular. In this way the notion of "the whole" achieves some sort of possibility. But to make wholes is also to make parts. It is to allow—indeed to call for—magnification. Better, it is to imagine magnification as a possibility that gives it some sort of sense. This is what makes it possible to say that Cyril Musgrave's narrative is a "detail" of the "larger" narrative with which I started this chapter. This would make it possible to locate Musgrave's narrative, from the point of view of "the whole story," in a footnote. To put it in a black box. Like the subsections of the brochure in the table of contents, it makes it possible to say that his story is a "detail." But which also, and in the same movement, makes performative the distributions of size. Including the performance of my "larger" narrative as, indeed, "the story as a whole."

Arborescence is a hierarchical structure, a system of points and positions, an axis—I am simply using Claire Parnet's metaphors—that domesticates particulars by locating them, without potential limit, in a salient and scaled space-time grid. It is a location that performs Haraway's god-eye trick, which makes a four-dimensional view from nowhere. It defines and performs the conditions of possibility for particular stories. And it is this potential to locate that shapes the nature of interferences between specifics. It also defines two great questions: the epistemological question, which asks whether this story is real, and the occluded ontological question, which asks about performativity as a function of interference, deferral, and the structures of possibility.

The tree is one of the great metaphors for disciplining interferences. "The whole world demands roots." I take it, however, that there are other relations and overlaps. That it is possible to perform other interferences.

Ending

In chapter 4 I suggested that "the project" is the performance of a cultural bias in favor of continuity. Discontinuities and stutterings, I there suggested, are effaced in a continuing slippage or oscillation between three versions of continuity—genealogy, system, and interests, or more generally, depth. This chapter re-presents this cultural bias in favor of continuity in somewhat different terms, by asking what happens if we consider the TSR2 project in arborescent terms. I have suggested that such a narrative is performative but performative in a particular way. For in addition to enacting a singular project by effacing the productive interferences of project multiplicity, it also performs what one might think of as ontological singularity. That is, it reenacts conditions of possibility that assume the objects in the world are contained in a singular Euclidean space-time box. That, in other words, the world is that way itself. Harshly, I have called such storytelling "collusive," though as I noted at the beginning of the chapter, in saying this I am not implying personal bad faith.

So what happens if we do not "collude"? If we resist the smoothnesses of such "modern" storytelling about projects, technological and otherwise, then the singularity of the world—its Euclidean space-time box with its time and its dimensions—begins to dissolve. We discover that chronologie(s) and scale(s) is (or are) themselves an

The Virtual Object

Either there is an object, out there, and then there are reports on that object, perspectives, more or less different perspectives, more or less satisfactory perspectives. In which case the aim is to understand how perspectives are built, and to look for good as opposed to bad perspectives in whichever way one prefers to make that distinction.

Or there are practices, narratives, and relations. Practices, narratives, and relations that perform subjects and objects into being. That *make* the knower and the known. Or better, that make multiple knowers and multiple objects of knowledge. In which case it becomes important to make sense of the interferences, overlaps, and methods of coordination, if such there be. The relations between the different objects. And between the different subjects.

We have seen that the first option takes us to epistemology, while the second draws us to ontology, to questions of making real, of realizing. It leads us to questions of performativity, to levels and conditions for performativity.

Is it a condition for performativity to imagine that there is, indeed, a singular object out there, separate from the knowing subject? Is it a condition for performativity to project a virtual object, a virtual reality, behind the appearance? Virtual not in the sense that it is recognized as a simulation but rather in the sense that pins appear, virtual pins, as in the school optics experiments carried out with mirrors and glass blocks: something out there, behind, beyond, that looks quite real. Appears to be singular. But is in fact produced. An effect. And multiple.

Perhaps the TSR2 project is a virtual object, and the decision to cancel the aircraft is a virtual decision (see Hennion 1996, Law 1998a, and Mol 2002a).

achievement or a performance rather than something given in the order of things. We discover, in short, that "the project" (the TSR2 project, the academic or policy project of studying the TSR2 project, the project of technoscience, the "modern project") participates in the performance of singular conditions that precisely renders it performatively possible. While there are, of course, alternatives.

I would like us to explore some of those alternatives. To make the oscillations between singularity and multiplicity available for inspection.

Splitting, not being, is the privileged image for feminist epistemologies of scientific knowledge. "Splitting" in this context should be about heterogeneous multiplicities that are simultaneously necessary and incapable of being squashed into isomorphic slots or cumulative lists. — Donna Haraway, "Situated Knowledges: The Science Question in Feminism and the Privilege of Partial Perspective," in *Simians, Cyborgs and Women: the Reinvention of Nature*

The strange history I have amassed for you to peruse is as much fable as history, yet always factual. — Michael Taussig, *Mimesis and Alterity: A Particular History of the Senses*

These fractal graphics could describe the patterning of maps or genealogies, but they would be maps without centers and genealogies without generations. It is the repetition, the not-quite replication, to which the viewer is compelled to attend. — Marilyn Strathern, *Partial Connections*

Narratives

Imagine a story. It's a story that narrates a series of events. It goes into detail, and it needs the details because they add to the story, provide evidence for it, help to explain what really happened. In short, a story like the kind of arborescent narrative with which I opened chapter 8.

Now imagine a pinboard. There are photos of friends or family members. Unpaid bills. Pictures and cartoons torn from magazines. Poems. Telephone numbers. Messages. Take-away menus. Newspaper articles.

A story and a pinboard do different jobs of work. They exist in different worlds. Crucially, they also help to *make* different kinds of worlds. And it is the making that is interesting, the performativity of storytelling on the one hand and pinboards on the other.

As we have seen, a story narrates. Events follow other events. Things are made to follow other things. This practice conforms to a tradition of continuity, of narrative, of telling how things arose, developed. And a powerful tradition it is, this tradition of ordering, of consistent explanation, of foundations, of origins. It's the tradition of the larger part of academic writing—history, sociology, technoscience studies. But also of many, perhaps most, parts of Euro-American culture: the modern novel; the nineteenth-century travel narrative; *Pilgrim's Progress; Beowulf;* the Bible story. As far as we can see in any direction we find a narrative tradition in the West that is more or less grand, more or less inclusive, more or less smooth, more or less foundational. A narrative tradition that is often, as Donna Haraway (1997) reminds us, also about redemption. Redemption is denouement.

And the pinboard? Its logic is quite different. It is one of juxtaposition, of pastiche, and yes, it claims that somehow or other these bits of paper, these postings, are more or less equivalent. That they exist together on the same surface and may expose themselves without the necessity of a single order, may jostle one another. Yes, they may make links with each other or overlap. They may resemble one another or differ. Help one another or not. But this is a logic that also supposes that they do not, except coincidentally, belong to one another as part of a single, larger whole. That they do not, except coincidentally, belong to the same narrative.[1]

How might we think about this difference, the difference between story and pinboard? How might we think about the difference between the work that each does? How might we think about their performative effects?

Let's start with narrative. Perhaps this comes in two versions. One is a grand narrative, smooth and modernist in style, which says, precisely, that we are at the end of the age of grand narrative. It says that storytelling is passing, and that an aesthetics of juxtaposition is coming. Making overall frames or contexts is thus on the way out, and local arrangements, resemblances, intensities, are on the way in. The story tells itself in many modes: philosophically or psychoanalytically in terms of shift from discourse to figure; in stories about the city, the overload of images, and the aesthetic performance of *flâneurie;* in accounts of the collapse of the distinction between sign systems and their referents and the consequent aestheticization of the world; in technologically determinist modes with stories about the rise of the visual, the rise of the television or MTV, and the decline of the text; and in more specific stories, for instance, about the rearrangement of subjectivities in personal computing brought about by the shift from the impenetrable depths and linearities of DOS to the surface transparencies and iconic relations of the Macintosh or Windows.[2] This, then, is a grand narrative—or a family of grand narratives.

The second version of storytelling is another family of more or less modernist grand narratives. It also tells that narrative is not over, not yet, or perhaps not at all. It says that the novel, or history, or progress, or the Enlightenment project is not dead. In its predominant sociological versions—those of reflexive high modernity—it tells that, yes, there are changes, that life moves more quickly, that there is constant churn. And then—pausing to observe that this was characteristic not simply of postmodernity but also of modernity—it goes on to argue that the stories we tell need to adapt themselves to modern times. Narratives shift; there are no firm foundations. But there are nevertheless temporary coherences, revisable stories that will hold contingencies together for a time, which means that the art of contemporary living and knowing is to make those narratives—and then to reflect upon them, adapting them as circumstances change.[3]

So we have two families of (more or less) grand narratives. They are explanations for change, for the limitations of grand narrative, explanations for the rise of the pinboard, the *flâneur,* or MTV. They are explanations for the "postmodern" logics of resemblance, overlap, association, for the postmodern aestheticization of the world. But they themselves are more or less "modernist" in style. And they are explanations in narrative form, more or less singular, more or less able to draw what is relevant together, to *render* it coherent and singular, to perform singularity. Discourse rather than figure (Lyotard 1984a; 1985).

But what of an alternative, the pinboard? In the academy we usually find that the pinboard is something to be narrated, or narrated away. In social theory or in technoscience studies we do not see, not much, *the performance of an academic version of the pinboard, the performance of resemblance, of collage or pastiche.* A notable exception would be the remarkable study of health economics by Malcolm Ashmore, Michael Mulkay, and Trevor Pinch (1989).

Here are some of the questions that would arise if we went the way of the pinboard:

—What are *possibilities* of juxtaposition? What does it do? What does it make?

—What might it be like to *be* an academic in a world of "postmodernity"? To *be* a "postmodern" academic? Or perhaps better, since the term *postmodern* is better avoided, one who was "nonmodern"?

—What would it be to know about things *indirectly,* perhaps allegorically, in a manner that does not aspire to center itself or, if it does center itself, does not seek to erase its multiplicities as it also performs singularities?

—What would it be to know, instead, *in tension* and reconcile ourselves, yes, to the loss of conclusions?

—What would count as *rigor* in the mode of the pinboard?

—What would be the *effect* of telling about things, of knowing them, in little patches? What would be the effects of juxtaposition?

—What kinds of knowing and knowing subjects would the logic of the pinboard perform into being?

The answer to these questions is that we do not know. And curiously, they are also questions that seem to be rather far down on most academic agendas.[4]

For yes, the loss of grand narrative is a loss, loss of stories with beginnings, middles, and ends; stories that cover the ground more or less smoothly and tell us how things hang together, explaining what lies behind the confusion of appearance. The redemption that lies after struggle with disorder—or other orders—in the Judeo-Christian tradition. These are losses. But with these losses there are also gains. The hope, then, is that by giving up big narratives that seek to order the whole, one might know in ways that are still rigorous but know differently. Not in a "postmodern" style, not, that is, as a celebration of fragmentation, the breaking of a modernist whole. But perhaps, nevertheless, in a post-Enlightenment mode, a pinboard mode. To know in a way that moves from one narrative to another, from one pictorial component to another, in order to secure other knowing effects.

This is what I have attempted in this book. I have tried to make the first juxtapositions in my own version of such a pinboard project. To see what happens if one makes a book of knowing in a pinboard mode. If one performs an academic pinboard.

There was a moment when, following Ashmore, Mulkay, and Pinch, I imagined one of the chapters of this book in the form of a pinboard. There are plenty of visual materials to do with the TSR2. Brochures, technical drawings, photographs, maps, sketches, flow charts, chronologies, doodles, graphs, instructions, working diagrams, bar charts, tables, cartoons, caricatures, films, radar scans, circuit diagrams, organizational charts. Public and private, personal and impersonal, human and nonhuman, views from somewhere and views from nowhere, even after the destruction of the TSR2 the visual materials are endless. And that says nothing of all the texts. Captions, snippets, articles, letters to the editor, press releases, policy documents, newspaper editorials, chapters, books, technical reports, personal letters. On and on and on. Juxtaposed materials, visual, graphical, textual. Such a chapter would have *performed* the resemblances and similitudes of an academic pinboard rather than *saying* it.

The book contains no such chapter. There are practical reasons, difficulty and the working of the Official Secrets Act being among them. Nevertheless, the book is written in a way that has much to do with the logic of the pinboard. Though individual chapters are not primarily visual in form, in the narrations I have often tried remake a version of this visual logic by working in a more or less broken or juxtapositionary mode. Or in some cases in a more linear manner, by describing the effects of juxtaposition as something other than a loss.

Chapter 2 performs a principled naïveté. It asks, What happens if you forget that this object, this TSR2 aircraft, is a single aircraft? The answer is that is you end up with a whole lot of different aircraft. A pinboard full of aircraft. And you end up with what Annemarie Mol calls the "problem of difference."

Why inflict this complication on ourselves? Because it raises an empirical question. If it's right, we are then faced with a problem that we never saw before. We have to work out how these different aircraft coordinate. How they turn themselves from many, back into one. If indeed they do.

Chapter 3 tells of a struggle between the author and the aircraft, this TSR2. It tells about the "obviousness" of the aircraft, and how that obviousness changed from time to time. "Obviously" an attractive piece of machinery, it was also and "obviously" a killing machine, "obviously" a good topic for a study in the social shaping of technology, and equally "obviously" an episode in need of policy recommendations.

It becomes apparent that there are too many "obviouslies." And if there are too many then none of them is obvious. Basic sociology, from Georg Simmel onward. Various forms of knowing. An enactment of the corroding logic of interdiscursive juxtaposition.

And the book as a whole is no different. A set of partially connected narrations, it is also in some measure a pastiche. It is an attempt to *perform* decentered or allegorical knowing. It is an attempt to edge toward and perform a set of alternative academic sensibilities having to do with association, resemblance, or similitude—and with multiplicity and difference. It is also an attempt to come to terms with an uncomfortable lack of completion—with the absence of redemption.

Accepting incompleteness is indeed uncomfortable because it presses against the conditions of academic possibility and its narrative performances. But—and this is why I think it is worth the discomfort of breaking the rules—*it is also to refuse to buy into object singularity.* Instead it imagines objects—and the worlds in which they subsist—to be *fractionally coherent.* Oscillatory between singularity and multiplicity.

Such is the potential payoff. And it is why this exercise is more than some kind of aesthetic game.

Here is the argument: juxtaposing "images" and making pastiches raises the possibility that the world is not a singular place. It raises the possibility that objects in the world, for instance an object like an aircraft, is, are, both multiple and singular. It raises the prospect that what appears to be one may also be many, and that coordination produces singularity. It also suggests that it is interesting, impor-

> Chapter 4 makes smooth stories about the design of the TSR2. There
> is a story about origins and evolution, a story about systems, and a
> story about the way in which technologies are shaped by social inter-
> ests. These stories are derived from the stories told by participants and
> close observers. But they are also the standard forms of narration in
> sociology, history, or technoscience studies. So chapter 4 is also about
> the "cultural bias" in favor of narration and the corresponding aversion
> to the logic of the pinboard.
>
> But why not simply embrace smoothness? Because it centers the ob-
> ject, rendering its noncoherences difficult or impossible to know. And
> simultaneously, the same point back to front, *it centers knowing*.

*tant, indeed vital to study that coordination, the ordering logics of
the fractionally coherent object.*[5]

This is the first reason for opting for the logic of the pinboard. Objects in the world are similarly assembled; they are more than one and
less than many. Though they narrate themselves otherwise, eliding
their multiplicities and putting them under erasure, *they are like a
pinboard in form too.* This is an interesting and important discovery.

But then a question. Why would this be a *discovery*? Why is it not
obvious? Why do we not constantly challenge the idea that the world
is a single, more or less coherent container that holds definite and
more or less narratable objects and processes? Why do we not accept the idea that it is somewhere between singularity and plurality?
I think the answer is that a strategy for knowing which rests on larger
narrative finds this difficult to say because of a performative contradiction. It can, of course, *say* it in as many words, directly: "The world
and its objects are more than one and less than many." It can say that
they are fractional, but when it does so, a larger narrative performs a
contradiction. For as it *tells* about fractionality, it also *performs* singularity. The problem, then, is that singular narrative presumes singularity. Fractionality may be told, but at the same time it lies beyond
the limits of its conditions of narration. It is erased. The logic of the

> If the possibility of decentered knowing or subjects is somewhat less
> of a scandal than the prospect of decentered objects, this is because
> there is a well-worn argument about the "decentering of the subject."
>
> If it remains a scandal at all, it is because decentered subjects and
> decentered forms of knowing are rarely celebrated within the academy.
> Though I'd argue that this means we are in a state of denial—or
> specific emphasis—which is also an aporia. This is one way of read-
> ing chapter 5 and its argument about "hidden heterogeneities." For
> what could be more centered than the subject position made by a
> single engineering formalism? If the formalism is centered, if it is self-
> present, then it is because it also denies its absences. Absence and
> presence, self-identity and Otherness, these make each other together.

pinboard, however, works the other way round. It precisely helps to
create the scandal of the object that is more than one and less than
many.

So this is the first scandal of the pinboard, the first place at which
it pushes up against and resists the conditions of narrative possibility
by refusing redemption into singularity.

The second scandal is closely related to the first, and it has to do
with knowledge, with what it is to know things. For—and this is a
truth universally acknowledged—there are different stories in the
world. Many of them.

But what happens if we pin them up, side by side, and look at them
all together? The answer is that sometimes they fit together. "This
happened, then this, and then this." In this way a single narrative is
restored, one with an overall point of view, a modernist smoothness.
Then the tension drains away. Matters are resolved. But sometimes
they don't fit—which is perhaps the more common predicament of
the storyteller. Then comes the narrative task of explaining differences, explaining them away, for instance as the product of faulty
vision, biased perspective, inadequate information, local interests,
whatever else.

Text and pictures. On the pinboard the two are held apart, but they are also juxtaposed. Sometimes the juxtapositions are artful. And sometimes they interfere with one another to produce striking effects.

In chapter 6 I talked about a sales brochure. The text is full of facts. It describes the aircraft, its attributes, its systems, its construction. There is everything that the rational, disembodied, strategic decision maker would need to know when thinking about whether he wanted TSR2 for his air force.

Alongside the text there are illustrations. Some demonstrate technical systems. But others appear to stress the speed of the aircraft, the thrill of flying it low and fast. Or to demonstrate the painless destruction of factories or bridges. Yet others do more complicated work concerning its invisibility, its propensity to hide, to be protected, indeed, to be nurtured.

In the chapter I make arguments about gendering work. But what's important here is that it is not possible, in the brochure, to *talk* about gender. Instead, text and pictures interfere with one another in gender-relevant ways. Both say that this aircraft is a powerful creature, but they do so in very different ways. Indeed, they do so in ways that are inconsistent or impermissible. It is not permissible to *say* that this is one hell of an aircraft to fly. It is not permissible to say that the aircraft is a powerful male icon. It is not possible to tell that it also partakes of the power of the woman in the patriarchal imagination. But it is possible to *draw* all these things.

This is the stock-in-trade of the academy, turning the mess on the pinboard into a story. But what happens if we refuse this? I have said that it makes the scandal of the fractional object. But it also leaves us with bitter fragments or, yes, with pastiche, with sets of similarities and differences, sets of interferences, with elements that do not hold altogether—or which hold together in one way or another but without an overall logic. Such is the corroding logic of juxtaposition.

To juxtapose "images" or elements of different narrations, to make pastiches, and then to look for different links between them raises the possibility that knowing is not simply—and should not necessarily simply aspire to be—a singular activity. It raises the prospect that there are different and valid knowledges that can be neither entirely reconciled nor dismissed, and suggests that knowing is or might properly be, a process that is also decentered, distributed, but also partially connected. The logic of juxtaposition renders it inappropriate, even impossible, to draw things together into singularity.

This is the second reason for pressing the logic of the pinboard. The *subjects* of the world—including the academic subjects of the world—are, or might be, similarly assembled. This possibility is interesting and important because *it requires fractional subjects to know, to apprehend, fractional objects.* Oscillatory subjects to know and apprehend oscillatory objects. It demands ways of knowing that escape the possibilities of single narrative, a single narrative that performs denial of their conditions of possibility.

So the logic of the pinboard may also be treated as a *methodology.*

Collusion: smooth narrative makes more or less smooth objects. With beginnings and middles and ends. And if we choose to talk about "technological projects" then we collude in performing these, in one way or another, as trajectories that began, grow, are shaped by this or that technical or social factor, and then end somehow or other, happily or otherwise. We perform "projectness," and with that projectness we collude to perform a world fit for projects—a singular time-space box defined by the ticking of the clock, the three-dimensional spatialities of the Euclidean world, and a set of objects in that world, some big, some small, some earlier, some later. Chapter 8 asks what happens if we don't do this. If we don't collude to perform a singular world. If we imagine making fractionalities. Or oscillatory worlds.

It is a methodology that makes it possible to know about features of the world that deny themselves when everything is drawn together into a single story.

Politics

If the pinboard makes complexity in a way that narrative does not, then this book makes complexity in a way that a historical or techno-science "story of a technology" would not. That's it. It *makes* complexity. For though the complexity of the pinboard was always there alongside the linear travel of storytelling, and the purificatory narratives of modernity were always both dangerous and dangerously misleading, to narrate and to juxtapose have an effect. Narration is not separate from what it narrates. And neither is the pinboard. Both are performative. So making something like a pinboard helps to perform pinboard objects into being. More or less. It talks down the eliding performances of consistent narrative. More or less. It performs objects, more overtly, as collages or pastiches, as multiplicities or fractionalities that escape the possibilities of singular narrative.[6] In short, it makes a difference to that mirror-image pair, modernism and postmodernism, and the dualist politics that they imply—the choice between a smooth ordering or a set of disconnected fragments. Making a pinboard escapes those binarisms and practices something that might be imagined as an *ontological politics*—a politics about what there is in the world (Mol 1999 and Law 1996b). What there might be in the world. An interference in the conditions of possibility for the kinds of things that might exist in the world. Between the singular and the plural. Homage, then, to Michel Foucault for teaching us that there is no innocent knowing. For to know is to perform, to participate. To make a difference, one way or another.

So far, so good. But now I want to deal with the question of political responsibility and irresponsibility. What is the character of this problem? It is sometimes said that to work with something like the logic of the pinboard is to simply to play at being "postmodern." It is to refuse to take matters seriously. No doubt the argument comes in various forms. There are questions to do with rigor, methodological questions. There are issues arising in theory to do with what methodology looks like if simple narratives and conclusions are not possible. And then there are ethical or political questions. For the suggestion—

It is a trope of the poststructuralists to talk of "play." "Just gaming" (Lyotard and Thébaud 1985).

How to think about this? As an Other to that which is serious? Perhaps. But then remember that to talk of play, for instance in an engineering structure, is to refer to looseness, capacity to shift, in what may otherwise appear to be a rigid structure. In which case "too much" play may be undesirable. But insufficient play is likely to be disastrous.

This suggests that a single structure is made possible because it is, at the same time, not a single structure but an assemblage of partially connected pieces. More than one and less than many.

no doubt it is sometimes an accusation—is that this kind of exercise is a form of politically irresponsible play. Or, again, that it merely indulges in a form of aesthetics. To make pastiches may feel good. It may even look good. But it is not really serious. And, in particular, it doesn't engage with the real political problems of the world.

To specify, the criticism is that if we don't tell powerful stories about the great distributions—for instance ethnicity, gender, or class—then we stand precious little chance of making a difference to those distributions. Instead we are simply playing from a privileged position.

No doubt there is something in all this. Pinboards don't work in the same way as stories, and politics is no doubt often well pursued through essentialisms—strategic essentialisms?—made in, and helping to make, larger stories. But there is an alternative view. It is that large stories, with their requirement for overall coherence, miss out on important features of the world. They miss out on the oscillation between singularity and multiplicity, on fractionality, which I have already discussed. But they also miss out on distribution.

Perhaps an analogy will help. In 1980 three sociologists, Nick Abercrombie, Stephen Hill, and Bryan Turner published a book called *The Dominant Ideology Thesis* (1980) in which they explored the Marxist use of the term *ideology*. They argued that society in its various antagonistic modes was not, as many Marxists believed, held together by a dominant ideology. If there were a dominant ideology, it was rarely important in keeping working people in order; at most it tended to help hold dominant groups together by securing certain attitudes to property—though even this was uncertain, at least in modern times.

If Abercrombie, Hill, and Turner were wrestling with the dominant ideology thesis, then perhaps what we are wrestling with here is something that is even more powerful. Perhaps we are running up against what we might think of as the *dominant narrative thesis.* I've already made this argument in other terms: the dominant narrative thesis proposes that behind appearances is a narratable unity. It also supposes—in its political or ethical forms—that unless we narrate that unity we are simply playing games while the powerless continue to suffer. But while there is no doubt game playing, the accusation of moral or political indifference is wrong. Indeed, it is dangerously wrong because the great distributions are sustained as much in the complex and fractional logics of pastiche as they are in the coherent narrations and processes of inequalities. And if this is indeed the case, then it is vital to explore the logic of the pinboard if we want to understand even quite conventional political inequalities instead of colluding with them by performing them again in their splendid singularity.

The great stories about distribution and inequality narrate coherences, for instance, in the forms of global capitalism, patriarchy, and ethnicity. Indeed it is precisely their capacity to relate deep coherences that turns them into great stories. But in the last twenty years we have also learned to be cautious. Zygmunt Bauman (1989) has taught us that the grand narratives of what we call "modernity" are ambivalent: to garden is also to make "weeds." But now we need to imagine something more, something yet more difficult. This is the possibility that inequalities may be sustained in the play of noncoherent interferences. In what I want to call obdurate incoherences.

And this is a third scandal, which after the substantive and the methodological scandals, we might think of as the political scandal: many inequalities and distributions are fractional effects of noncoherence. Because of this they cannot be addressed and captured in single stories but can only be addressed through a series of different stories that don't add up very well. Or, if they do, they do so indirectly, as it were allegorically, through the logic of the pinboard. And here is

What is it to make a political decision?

In chapter 7 I deconstructed the decision to cancel the TSR2. It would be better to use quotes: I deconstructed "the decision" to cancel the TSR2. For it turns out that this decision was several different decisions. Each of the relevant cabinet ministers was making his own decision. While each at the same time believed that he was making a single, collective decision. In putting it this way, I am refusing the "dominant narrative thesis" which would insist that there was indeed a single decision—and then a set of misunderstandings or different perspectives. Two points about this.

First, the dominant narrative thesis is performative. The commitment by cabinet ministers—and everyone else—to the idea that they were making a single decision meant that they were, indeed, able to take "a decision."

Second, and more important in the context of the politics of interference, it is the *differences* between the decisions, the fact that it takes (at the very least) some effort to order them into a dominant narrative, that secures the strength of "the decision" in the first place. This is what I mean if I talk about "obdurate interference." If it had all been narratable as consistent, then the cabinet ministers would never have been able to make their "decision" in the first place.

where I want to end this account, to stop the juxtapositions that I have made on my version of the uncomfortable pinboard.

What I believe is something like this. TSR2 was an immensely obdurate object, a huge modern military project, one that employed tens or hundreds of thousands of people. It was, as we know, eventually destabilized—and the consequence was a conventional political scandal of considerable proportions. I've noted that as I worked my way into the project and collected material I was charged, in one way or another, with the responsibility of narrating the project and

In *Irreductions* Bruno Latour tells a story about colonialism. The merchants hated the bureaucrats. The soldiers were contemptuous of the cartographers. The priests despised the hacienda owners and the engineers. But this noncoherence made colonialism stronger rather than weaker. Atheism, the collapse of the markets, the failure of the bureaucracy, no particular breakdown necessarily played itself through to undermine colonialism "as a whole." There was no Leninist "weakest link." Instead there were partial connections which operated to perform an established disorder in an oscillation between singularity and multiplicity.

its downfall. "Why," I was being asked, "did it go wrong? And what can we learn?" This was uncomfortable. But I now believe that it was uncomfortable not only for the obvious political reasons but also because it implied the need for a single narrative, a dominant narration.

If we turn the question around and ask, rather, how it was that the project managed to hang together for as long as it did, then *the answer cannot be narrated in a single story at all.* Yes, there are stories, many stories about how it held together. Mostly top-down, managerialist stories, stories about control, ordering. I don't doubt that those stories tell us something important. There was plenty of narratable control and ordering. But neither do I doubt that they miss something. They don't, or so I am arguing, simply miss out because they are incomplete—though no doubt this is always the case. They also miss out because *the project was held together by interferences between the narratives that cannot be properly narrated within those narratives themselves.*

This is the point of talking about "obdurate interferences" or "obdurate incoherences" and why it is important to mobilize a metaphor like that of the pinboard. For it is the case, or so I am asserting, that the distributions of the social world—project distributions, political distributions, but also the more classic distributions of ethnicity, gender, or class—are sustained as much in narrative incoherence as they are in narrative coherence. Are sustained as much in interference between multiplicities as they are in successful and singular enactment. Thus the need—the great need—is to create sensibilities and toolkits that will allow us to sense, to work upon, and to interfere with those distributions once they escape the possibilities of single stories and enter the logic of oscillation.

There are, of course, many straws in the wind. Old stories about the ways in which reeds bend before the hurricane that destroys mighty oaks. Newer stories about the fluidities of successful social and technical forms, the ways in which these change themselves, never standing still long enough to draw their boundaries or narrate themselves into a single structure. There are the metaphors of fractionality, of partial connection and Donna Haraway's accounts of the current disorder. So there are many straws in the wind. And this is, indeed, encouraging. For this is where I stop this particular story. In the firm belief that a concern with noncoherence, the logic of the pinboard, the play of the fractional, which is more than one but less than many, is not some irrelevant aberration. It is not simply a game invented by intellectuals to make their discourses yet more complicated and inaccessible. Rather, it is the invention—or the reaffirmation—of ways of knowing that are simultaneously modest and complex, ways of knowing that are necessary if we are to imagine and interfere successfully in the workings of the current disorder. To imagine ways between the singularities and multiplicities of modernism and postmodernism.

The era of the pinboard is upon us. Our simplicities will no longer serve.

1. Introduction

1 At least in the more optimistic version, as advocated by Anthony Giddens (1990).

2 See Bauman's account of the Holocaust (1989).

3 In the distinctive way in which I will use it, the term *singularity* thus stands in contrast to *plurality,* rather than referring to something that is remarkable, extraordinary, or unique.

4 Pictures of fractal Mandelbrot curves have entered the Euro-American consciousness. These are lines that elaborate indefinitely as magnification is increased—more folds, curves, and bends appear each time the scale is increased. What starts as a line in one dimension thus starts to occupy more and more of a second dimension, the surface on which it is drawn. Hence, more than one dimension, but less than two. For discussion in the context of the social sciences, see, in particular, Marilyn Strathern 1991 and 1992, and Roy Wagner 1991.

5 This is one of the roots of the political objections to postmodernity: that it appears to be a new way of securing privilege under the guise of pluralism. I return to this issue in the final chapter.

6 The term *drawing things together* derives from Bruno Latour's essay (1990).

7 Those scholars whose work I know best include Michel Callon (1986a, 1986b; Callon and Law 1995), Kevin Hetherington (1997a, 1997b, 1997c, 1998, 1999; Hetherington and Lee 2000), Donna Haraway (1991b, 1992, 1997), Bruno Latour (1993, 1996a, 1996b), Annemarie Mol (1998a, 1998b, 2002a, 2002b; Mol and Berg 1994; Mol and Law 1994, 2001; Mol and Mesman 1996), Ingunn Moser (1998, 2000; Moser and Law 1998, 1999), Vicky Singleton (1993, 1996, 1998; Singleton and Michael 1993; Law and Singleton 2000a, 2000b, 2000c), and Marilyn Strathern (1991, 1992, 1996, 1999). Then there is the work of the philosophers, including Gilles Deleuze and Félix Guattari (1998), Jacques Derrida (1978), and Jean-François Lyotard (1984, 1991), and the allegorists, prominent among whom one would count the more or less rediscovered writing of Walter Benjamin (1985, 1992).

8 The exceptions include certain work in feminist traditions—including, in particular, attempts to understand the performative and situated character of knowing by Carol Cohn (1987, 1993), Donna Haraway (1991d, 1997), and Sharon Traweek (1988a, 1992, 1995a, 1995b)—and a tradition of reflexive sociology of science that seems to have received relatively little attention in recent years. For examples of the latter see Steve Woolgar 1988b; his edited collection 1988a, which includes papers by Malcolm Ashmore, Bruno Latour, and Trevor Pinch; Ashmore, Mulkay, and Pinch's book-length study of health economics (1989); Ashmore's *Reflexive Thesis* (1989); and my own attempts to explore the situated character of knowing managerial practice (Law 1994).

9 See the argument laid out in Pickering 1993, and developed at book length in Pickering 1995.

10 For discussion in the context of actor-network theory, see Law 1999.

11 Most of these metaphors are drawn from Deleuze and Guattari 1988.

12 For discussion of distribution in the context of technoscience studies, see Law 1991a.

2. Objects

1 In the semiotics I am implicitly using here, it would be appropriate to say that the different exhibits I have cited perform a series of different subject position, object position pairs.

2 This assumption is not unlike that which underpinned Harold Garfinkel's program of ethnomethodology—a method for throwing taken for granted assumptions into relief. See Garfinkel 1967.

3 This problem is described in a series of publications. See Mol 1998b, 2002a; Mol and Berg 1998; and Mol and Elsman 1996.

4 Syntax does not necessarily carry the day. There were those who thought, both for technical and strategic reasons, that it was not possible to combine all these roles within a single aircraft. See, for instance, Zuckerman 1988. I will return to this point later.

5 The term, and its contrast, the rhizome, is developed at length in Deleuze and Guattari 1988. See also chapter 8 (this volume), where I explore the arborescent character of technoscience narrative.

6 It thus works like Jasper Johns's "target" paintings. See Fisher 1991. Note, in addition, that there are further classes of deferrals. For instance, material deferrals that would take us outside the text or beyond pictorial and cartographic representation, if we were able to follow, for instance, into sets of relations in metal and plastic, flesh and blood, and into what they call the "real aircraft." Into materials that are, as we might say, absent but also present (see chapter 5): into textural rather than textual "representation." So the brochure works to make a unitary object, if it works, in part because it also rests upon *material deferral.* Let's state the obvious. The reader/subject is not being performed as a position that takes the brochure itself to be a weapon. The reader is being constructed, instead, to understand that the brochure *describes* a weapon. That it, the brochure, represents something in another material modality. This, then, is material deferral, and it is central to what one might think of as the narrative potential of the brochure. Though actually it is a little more complicated because, in 1962, there was no TSR2 aircraft. The possibilities of material deferral are limited. The brochure has to report promises. Expectations. Descriptions of the Operational Requirement. This, then is another form of deferral. Into the future. One that is, not coincidentally, well known in high technology.

7 For exemplary studies of cartography, see the work of David Turnbull (1993b, 1996, 2000).

8 It is possible to argue that all forms of depiction, including perspectival

representation, are working surfaces. For the argument developed in the context of still life see Fisher 1991. For a discussion of a somewhat similar position developed in phenomenology see Jay 1993, 314–15.

9 The approach is most associated with the work of Thomas P. Hughes (1979, 1983). See also, however, Latour 1988b.

10 No doubt it is possible to tell stories about high reflexive modernity here, stories that tell about feedback and the self-monitoring of cybernetic systems.

11 In Latour's semiotic understanding, there is no place at all outside the narrow networks (he uses the metaphor of termites and their galleries). If we were to turn this semiotics into systems theory, we would have to say that it is a "closed system" (though the implications of Latour's approach are no doubt very different). In Hughes's understanding, systems are "open." Thus it is not that there is nothing outside the system—an environment. It is rather that, in a characteristic trope of high modernity, significant proportions of that environment are absorbed and regulated by the systems builder.

12 This points to a large literature on reflexive high modernity. In the context of technoscience studies, see the work of Karin D. Knorr-Cetina (1992, 1996).

13 There are also other things going on. One possibility is that the reader is being made to understand that the aircraft will negotiate a safe passage to this vanishing point because it is invulnerable. And it is invulnerable because it is like a cocoon, wrapped up in a different world. Thus the picture represents an aircraft world that is close to the ground—for we see the ground "moving"—but a world that is also far removed from that ground. If we were to make the argument textually we might say that proximity to the ground is a sign of invulnerability because it gives little chance for the launch of surface-to-air missiles. But the visual "argument" being made is more than that. It has to do with the way in which the subject-viewer is so close to, and yet so far from, a static world of mundanity. Juxtaposed with mundanity, it is also utterly *removed* from the mundane. Which, or so I take it, is also a reading of exhibit 2.1: the aircraft is on its way, detaching itself from the viewpoint of the subject, as it departs into the great blue yonder. I shall return to these themes and explore them in more detail in chapter 6.

14 Two examples relevant to technoscience studies are Stone 1995c and Turkle 1996. See also chapter 3 in this book.

15 It is also, however, in part a metaphysical question.

16 I am referring here to work by Annemarie Mol on arteriosclerosis (Mol 1998b, 2002a; Mol and Berg 1994; Mol and Elsman 1996; Mol and Mesman 1996); Vicky Singleton on the UK Cervical Screening Programme and on advice about sudden infant death syndrome (Singleton 1996, 1998; Singleton and Michael 1993); John Law and Vicky Singleton on alcoholic liver disease (2000c); Marianne de Laet and Annemarie Mol on the Zimbabwe bush

pump (2000); Anni Dugdale on IUD advice (Dugdale 1999); and Charis Cussins on the practices in a fertility clinic (Cussins 1998a, 1998b).

17 There is a parallel between this and the ontological choreography so movingly described by Charis Cussins in her work on in vitro fertilization (1998a).

18 There is an obvious parallel between this and Annemarie Mol's case study of lower limb arteriosclerosis with its frequent disjunctions between a "single textbook disease" and the different practices for producing that singularity. See also, however, the case of alcoholic liver disease discussed in Law and Singleton 2000c.

19 Perhaps there is a parallel between this and the necessary incoherences in the UK Cervical Screening Programme described by Vicky Singleton. See Singleton and Michael 1993. The point is that the ability to combine these roles in the first place was one of the factors that moved the aircraft off the drawing board.

20 This position is developed in actor-network theory.

21 For further details on linear perspective see Alpers 1989, Dunning 1991, Edgerton 1976, and Gombrich 1989. On epistemology, perspectivalism, and ontology, see Law 1996b, Law and Benschop 1997, Mol 1998b and 2002a, and Strathern 1991.

3. Subjects

1 Much might be said about allegory, most notably that it does not find a ready place in the representational regimes of modern science and social science. Aside from a few exceptional practitioners—of whom no doubt Walter Benjamin is the most notable—it has, at least until recently disappeared from the representational practices of the academy. For further discussion see Hetherington and Law 1998, and for a beautiful contemporary example of the indirections of the allegorical, see Hetherington 1997c.

2 Having introduced the quotes to signal my reservations, I will now drop them.

3 Ironically, while I was working on this chapter in May 1996 it was announced that the last of these Polaris submarines was being decommissioned.

4 I have touched on this in the previous chapter, and will return to it in chapter 6. For a brief review, see Wajcman 1991 and 1998. In the context of military technology, see Cooke and Woollacott 1993.

5 The term *partial connections* is another of Donna Haraway's phrases (see 1991b, 181), one picked up and explored at length by Marilyn Strathern (1991).

6 For a beautiful discussion of literary witnessing, see Clifford Geertz 1988.

7 There is a large literature on this topic. For those who like historical stories, the creation of the space of the laboratory and the character of witness-

ing by gentlemen in the seventeenth century is described in two wonderful books by Steven Shapin and Simon Schaffer (1985) and Steven Shapin (1994), though see the commentary in Haraway 1997. For those who prefer ethnography I would recommend Sharon Traweek's *Beamtimes and Lifetimes* (1988a), where she considers the related way in which scientific heroes are constituted, heroes, that is, who have (in the mythology) wrested truths from an unwilling nature. Again, the iconography is gender saturated.

8 More gendering, to be sure. For many tropes of artistic heroism, see Wheelock 1995.

9 I learned this term from Sharon Traweek, whom I thank. She is often accused of such "vanity ethnography." In what follows I imply that such accusations presuppose a truth regime that misunderstands the character of such writing. For further discussion of the construction of regimes or economies of representation see Hetherington and Law 1998.

10 The material character of such practices has been described by David Turnbull (1993a) for the case of Gothic cathedrals.

11 For further reading on reflexivity, see Ashmore 1989; Ashmore, Mulkay, and Pinch 1989; Giddens 1990; Gusterson 1995a, 1995b, and 1998; Latour 1988c; Traweek 1995b; and Woolgar 1988a and 1988b.

12 This is the implied (though friendly) criticism made of his work by Zygmunt Bauman in his *Modernity and the Holocaust* (1989). For further commentary, see Theweleit 1987, volume 1.

13 I want, quite unfairly, to cite Walter Benjamin in this context. For a fine explanation as to why this is unfair, and an exposition of the arcades project, see Susan Buck-Morss 1991. But see also Kevin Hetherington, *The Badlands of Modernity* (1997a).

14 On this, see Paul Rabinow's extraordinary *French Modern* (1989), which may be read as a continuation of Michel Foucault's project by other means.

15 See Foucault's series of books on the construction of the sexuality of modernity, starting with Foucault 1981.

16 For discussion of the multiple spatialities of the architectural and the discursive, see Hetherington 1997c.

17 I am most grateful to Martin Gibbs for extensive and productive discussion about the analytical possibilities of the notion of interpellation.

18 See Althusser 1971. Note also that the term is used and explored by Teresa de Lauretis (1987).

19 For explication in the context of military and other gendered technologies, see Law 1996a and 1998b.

20 "What I am attempting to bring to light is the epistemological field, the *episteme* in which knowledge, envisaged apart from all criteria having reference to its rational value or to its objective forms, grounds its positivity and thereby manifests a history which is not that of its growing perfection, but rather that of its conditions of possibility; in this account what should ap-

pear are those configurations within the *space* of knowledges which have given rise to the diverse forms of empirical science. Such an enterprise is not so much a history, in the traditional meaning of that word, as an 'archaeology'" (Foucault 1970, xxii).

21 See Foucault 1986, and Hetherington 1997a and 1997c. I am most grateful to Kevin Hetherington for extensive discussion of the significance of the heterotopic.

22 Or even nonmodern. For discussion see Latour 1993.

23 A topic explored in both real and imagined conversations with Susan Leigh Star. For a trace of these conversations, see Law 1994.

24 The results of this work are reported in Law 1988 and 1992; and Law and Callon 1988, 1989, and 1992.

25 I am not endorsing empiricism by the back door. I am making, albeit very rapidly, an argument to do with intertextuality: the relationship between narratives, intertextualities that perform empiricist effects as a version of the god-eye trick.

26 The direction of parasitism is defined, tautologically, by defining this style of narrative as "esoteric." Its one-way character, however, is not given in the order of things—and esoteric forms of narrative have not infrequently emerged into and remodeled other story forms. Note how the "empiricism" of plain history may be understood as an effect of interdiscursive commerce—the performance of certain moves as "factual" and a ground upon which other moves may be made.

27 I am grateful to Maureen McNeil for illuminating discussion about the gendering of aesthetic narratives.

28 Beamont (1968, 152). Roland Beamont was the pilot primarily responsible for testing the TSR2 aircraft. He is here describing a moment on the first and only supersonic flight of TSR2 on February 22, 1965.

29 Academic and policy narratives need not interfere in a destructive or immobilizing fashion. Very often, for instance, in the form of stories about large technical systems, they do not. Interference as destruction or erosion is a contingency.

30 In Graham Dawson's suggestive study of British heroic masculinities (1994), he uses a Kleinian approach to explore the possible intersections between various incompatible Imagos generated in the place between introjection and projection, and considers different versions of composure including manic splitting and reparation.

31 Ten years later I returned to RAF Cosford. The erotics and aesthetics of that ambivalent day are described in Law 1996a. For further discussion of the ambivalent pleasure of machines see Brenna, Law, and Moser 1998, and for wider discussion of masculine heroism, see Dawson 1994.

32 On difference between specificities, and their links, see the exemplary work of Annemarie Mol who, with her collaborators, has explored the diversity and coordination of medical practices within certain parts of medi-

cine. See Mol 2002a; Mol and Berg 1994; Mol and Law 1994, 2001; Mol and
Elsman 1996; and Mol and Mesman 1996.

33 I have tried to survey some of these practices in Law 1998b.

4. Cultures

1 See Traweek 1988a, 1992, 1995a, and 1995b; Mol 1998a, 2002a; Mol and
Berg 1994; Berg and Mol 1998; Mol and Law 1994, 2001; and Mol and Mesman 1996.

2 The classic study is no doubt Anderson 1991, but see also, for instance,
Dawson 1994.

3 The history of these two aircraft and English Electric is described in Ransom and Fairclough 1987.

4 This story is described in, for instance, Ransom and Fairclough (1987).

5 English Electric/Short Bros. 1958, 1.S.2. The Viscount was a civil aircraft
built by another aircraft firm, Vickers, which we will encounter shortly.

6 English Electric/Short Bros. 1958, 1.S.4. Charles Gardner, the company historian of the British Aircraft Corporation, writes that "[the P.17A] was based
on a concept of initial simplicity to get the prototype aircraft airborne as
quickly as might be, and then for nav-attack and other such equipment . . .
to be added later" (Gardner 1981, 30).

7 On the character of origin stories and narratives of redemption, see Haraway 1989 and 1997.

8 De Haviland, the Hawker Group, Bristol Aircraft, and Blackburn, all of
these companies were in touch with government and the people working
for the Air Force Operational Requirement Branch. And they were working
away on their own designs.

9 Marilyn Strathern (1996) reminds us that the family tree and its related
questions of descent and pedigree perform themselves as an unremarked
method for organizing similarity and difference in many Euro-American
contexts that are far removed from the family.

10 Though this is not to say that it had been completely successful.

11 Rolland Munro has undertaken subtle studies of change within firms that
explore the construction of a past that is constituted as dépassé. I am grateful to him for discussion of this point. See the introduction to Mouritsen
and Munro 1996.

12 Note that in its brochure English Electric also talks of its "weapons systems" philosophy. See, for instance, English Electric/Short Bros. 1958,
1.S.3.

13 The first prototype of a fighter aircraft called the Hunter appeared in July
1951, but it was discovered that when the guns were fired at high altitude
the engine tended to stall. And, the White Paper noted, "A single intractable
aerodynamic problem from time to time monopolised the flying time of all
the available aircraft or grounded them so that all other development flying

. . . had to be put back. For instance for a long time difficulty was experienced in making the Hunter air-brake slow the aircraft effectively without upsetting the pilot's aim" (HMSO 1955, 6).

14 The term *cultural bias* comes from Mary Douglas (1982), though in the present instance it has nothing to do with her grid-group analysis. I use the term with some trepidation, partly for this reason, and partly because it might be misunderstood as endorsing the possibility that cultural bias might somehow be avoided. But there is, according to Donna Haraway among many others, no view from nowhere. We all have the privilege of partial perspective. The issue, then, is what we find, or choose to find, as unsatisfactory or unjust among the many partialities within which we are immersed.

15 *Lone Ranger:* "Tonto, we're surrounded." *Tonto:* "Whadya mean 'we,' white man?"

16 For "puzzle solving," see Thomas Kuhn 1970; for "technological frame," see Wiebe Bijker 1987 and 1992.

17 I place "abstract" within quotation marks to note that something strange is happening here: killing, or more precisely "lethality," is being displaced and in some sense turned into a disembodied game. There is a large literature on this phenomenon—see, for instance, Cohn 1987 and 1993, der Derian 1992, Rosenberg 1993, and Robins and Levidow 1995—which can be seen as an expression of modernism that, like semiotics, desacralizes established categories.

18 On this transmutation, see Roland Barthes 1977.

19 An often-noted feature, to be sure, of modernity and its ruthless market logics. "All that is solid melts into air."

20 The term *configuring the user* comes from Steve Woolgar 1991.

21 This story exhibits similarities that are certainly not merely coincidental with contemporary writing about the body and in particular the immune system. See, for instance, Martin 1994 and Haraway 1991a.

22 Large systems theory and various versions of actor-network theory with its ruthless (and one might add) "semiotic" approach to (the lack of sanctity of) specific elements come to mind. See Hughes 1979 and 1983, and Law and Callon 1989.

23 A point beautifully made in the context of actor-network theory by Nick Lee and Steve Brown (1994).

24 See the studies grouped together in Mackenzie and Wajcman 1985 and 1999, and particularly those by David Noble, Ruth Schwartz Cowan, and Cynthia Cockburn.

25 A fine and empirically relevant example is Donald MacKenzie's study of missile guidance systems (1990). For a more explicitly theoretical statement, see Barnes 1977.

26 A classic example, again relevant to military technology, is Sapolsky 1972.

27 Versions of this approach, explored for empirical material relevant to the present study, are to be found in Law and Callon 1988 and 1989.

28 The links between what I am calling "system" and Michel Foucault's modern episteme are obvious, though I have not here linked system building to the self-reflexivity that is of such interest not only to Foucault (1970) but also to many contemporary commentators such as Knorr-Cetina (1992).

29 For a study of the importance of qualitative difference in the sociology of knowledge, see Mannheim 1953.

30 The term *grand narrative,* of course, comes from Lyotard 1984, though I would also want to link it to Foucault's modern episteme (1970).

31 Here I draw on the metaphor used by Haraway (1991a) and by Strathern (1991). Haraway also talks of stuttering in 1991d.

32 I believe that Fredric Jameson is pointing in this direction in parts of 1991. See also Law 1997 and 1998a.

5. Heterogeneities

1 This, of course, is a poststructuralist trope, though also one found in feminist technoscience studies (Haraway 1997). It is, for instance, crucial to the writing of Jacques Derrida (1978) and those writers such as Robert Cooper who have explored poststructuralist tropes in particular locations. See, for instance, Cooper 1986, 1989a, and 1989b, and Law 1995.

2 The argument is developed in Latour 1993. Modernity, he says, claims the purity of a clear distinction between the human and the nonhuman, between the social and the technical. But at the same time, it productively generates all manner of hybrids, part human, part nonhuman. So modernity, as it were, denies itself in its claims to purity. Latour does not, I think, talk of oscillation, but there are partial connections between his argument and what I am suggesting here.

3 The figures are taken from a short brochure produced by English Electric (1959) for senior RAF and government personnel. The figures quoted differ marginally from the full-length brochure (English Electric/Short Bros. 1958), though the differences do not affect the argument.

4 See, for instance, Callon 1991. Actor-network theory can indeed be understood as a form of monadology in which each actor includes everything. The point is explored by Latour (1988a), and has been recently considered by Law (2000a). The condensations of monadology are further explored by Law and Singleton (2000a).

5 For commentary on *translation,* a term that derives from Michel Serres, see Serres 1974, 1992; Callon 1980, 1986b; Latour 1987, 1999a; and Law 1997, 1999.

6 The differences between a General Operational Requirement (GOR) and an Operational Requirement (OR) are not important in the present context.

7 For more on Otherness, see Hetherington 1997a, 1997c; Taussig 1993; Mol 1994; and Law and Mol 1998.

8 The argument is developed in technoscience studies by Donna Haraway (1991a). As in other STS studies, it should also be noted that the present analysis is impartial with respect to (what is sometimes called) truth and falsity. I am not saying whether the Russians "were" or "were not" an enemy.

9 English Electric/Short Bros. 1958, 1.S.6. Consider also this: "Only the most phlegmatic and unimaginative individual can fail to take a keen interest in the running of his only engine when he is a few hundred miles from the nearest land or the nearest area of population or line of communication" (Vickers Armstrong 1958a).

10 Vickers Armstrong 1958b, 2. This is a short, glossy version of the Vickers Armstrong submission in response to GOR 339.

11 See Jacques Derrida 1978 and also Fredric Jameson's discussion of movement in representation where he writes about the Westin Bonaventure Hotel (1991, 38–45).

12 For discussion of tension in a related context see Robert Cooper and John Law 1995.

13 David Turnbull (1993a) argues that some knowledge traditions are decentered. I am suggesting that centered practices are also, and necessarily, decentered.

14 As, for instance, Gilles Deleuze and Félix Guattari argue (1988), though they sometimes seem to lapse into a postmodern romanticism. For commentary on this, see Braidotti 1994 and Moser and Law 1999. See also Kevin Hetherington's reflection on the different character that heterogeneity has taken since the early modern period (1997b).

6. Aesthetics

1 Pluralism reenacts a version of singularity because it presupposes a single space that may be carved up into different regions.

2 Though interference does not necessarily produce (or oscillate in relation to) singularity. See chapter 3 (this volume), and Law and Singleton 2000c.

3 I am grateful to Maureen McNeil for pointing out that questions of aesthetics in a created opposition to technics are related to questions of gender in patriarchal oppositions between masculinity and femininity. Two points here: (1) the power performed by or attributed to technology does not necessarily line up and resonate with gendering tropes, and (2) the idea that "the aesthetic" is associated with the feminine is an interference that deserves exploration in its own right, though I do not attempt that in the present chapter.

4 I should also add that in the original this depiction is in color.

5 That scaling may be treated as an effect is implied in the discussion of the

table of contents of the brochure. See chapter 2. See also Callon and Latour 1981.

6 In the present context there is an immediate irony around the status of the "technical." This is more usually treated as a subordinate class that can safely be ignored or delegated in favor of something treated as more important—for instance politics or society. But here the technical is being contrasted favorably, or so I read it, with the "aesthetic."

7 This is explored at length in Mitchell 1994, but the argument about modernism and purity is developed in many places. See, for instance, Latour 1993 and 1996b, but cf. Elam 1997.

8 See Bryson 1990, 81. The distinction between the aesthetic and the functional is related to that between substance and form explored by Bourdieu in *Distinction* (1984).

9 For discussion in the context of art, see Fisher 1991.

10 The subject of an almost limitless literature. See, for instance, Edgerton 1976, and for a contrast with an alternative, Netherlandish tradition, see Alpers 1989.

11 All of which is, to be sure, most strongly gendered. For discussion of the construction of nature, culture, and gender see Jordanova 1980, together with other papers in MacCormack and Strathern 1980. Haraway explores the issue further in the context of technoscience studies in 1992, n. 14.

12 The aesthetics are here in tension with the technics. There is no need for the aircraft to fly low over "home" territory. It is only over enemy territory that this is necessary because it is only here that it might be attacked by radar-guided, surface-to-air missiles. The picture would be more "realistic" (why do we allow the "technical" to define what counts as realism?) if it depicted radar antennae and missile launchers, but doing this would undermine an "aesthetic" version of the distribution of in/vulnerability in favour of one posed in "technical" terms. Alternatively, it could be that the aircraft is flying across enemy territory, back toward the promise of home. Such a suggestion has been made by a number of those with whom I have discussed this visualization.

13 Despite the often-remarked gendering trope (Staudenmaier 1994), the present contrast is not exactly that of light and dark, for the houses and the trees below us are clear enough. It is just that the vanishing point is brighter.

14 For discussion of megalography (stories of heroism) and rhopography (everyday stories), see Bryson 1990.

15 The combination of words and visual representation is common in twentieth-century art. One example: the Georges Braque painting *Clarinet and Bottle of Rum on a Mantelpiece,* where the bottle of rum appears, at least in part, as the letters *Rhu.* And this is not the only piece of textual play that appears in this painting (see Tate Gallery T02318).

16 The Hollywood trope, to be sure, also hints at transcendence, the move

beyond the screen and the film, to the endless frontier performed in this seemingly endless version of the American imagination.

17 It also implies quick and decisive action, perhaps indeed an end to the enemy. For there is a further important "detail": the coincidence between the illumination of the vanishing point and the corner of the *S,* which stands for "strike." The aircraft, we now begin to see, will strike like a bolt of lightning, fast and furious, against an enemy reduced to passivity. Another distribution, another distribution of agency.

18 The relative locations of the vanishing point and the buildings is also interesting. The picture is, to be sure, centered on that vanishing point whereas the buildings are displaced to the periphery, squashed, as one might say, into the margins. Arguably, this implies an allocation of relative value (as is suggested by Philip Fisher [1991] in his study of Jasper Johns's "target" series). This may in turn be understood as a gendering trope, as it is by Griselda Pollock (1988) in her study of women impressionists (though in the case of the TSR2 brochure there is no sense that the compression of the domestic is experienced as a matter of discomfort). Note that Fisher assumes (perhaps rightly in the case of Johns) that value declines as one moves away from the center of the target. But this implies that there is, as it were, still the possibility of a calculable form of value. For Pollock, the women impressionists are dealing with a clash between public and private space, which stand in opposition to one another.

19 This raises questions about the relationship between the senses. "Ordinarily," writes Svetlana Alpers (1988, 24), "sight is necessarily out of touch because one must be a certain distance from what one views," though her work on Rembrandt, and indeed the seventeenth-century Dutch school in general, suggests an intimate relationship between the art of describing and the value of touch. Note that sight is often set up in contrast with touch, as unreciprocal, a form of distant domination, which position, or so Martin Jay (1993) argues, is a persistent and erring trope of French contemporary thought.

20 It is instructive to note the space between such semioticians as Louis Althusser and Bruno Latour in this respect. For Althusser, who was well versed in the decentering semiotics of Lacan (see Jay 1993) but also wrestled with the "last instances" of Marxism, differences were a problem: "This concert [of ideological state apparatuses] is dominated by a single score, occasionally disturbed by contradictions" (Althusser 1971, 146). By contrast, as we have seen, for Latour contradictions are not a disturbance but are rather a source of potential strength. See his "Interlude V" in *Irréductions* (Latour 1988a, 201–2). I explore this issue in chapter 3.

21 See also Cussins et al. 1992.

22 One of these is Thomas Kuhn's (1970) account of scientific practice (though Kuhn is certainly not abandoning theoretical humanism). And then my own (1994) use of Kuhn's vocational puzzle solving as a "mode of ordering"

which also, however, reflects Zygmunt Bauman's (1989) relative skepticism about the program of the Enlightenment.

23 It is possible to imagine nature (or indeed any set of objects or processes) as a set of "virtual objects": that is, objects that are projected as preexisting behind the practices in which they are represented. See Antoine Hennion 1996, John Law 1998a, and Annemarie Mol 2002a.

24 In common with the other citations and pictures, it therefore counts as a "scenario." See Michel Callon 1980.

25 This point might also be made of the landmarks discussed earlier. They don't count as a threat, but they have the capacity to act as resources because the aircraft is able to detect them with its Doppler radar. Another way of putting this, which is one of the modes of associating identified in chapter 2, would be to say that the landmarks have been colonized and incorporated within the network or the system.

26 I'm not certain about the specifics of the Lacanian account offered by Griselda Pollock (1988) in her analysis of pre-Raphaelite representations of women, but there is at least a superficial resemblance between the troubled pre-Raphaelite attempts to insist on "feminine" perfection and this brochure's insistence on the technical competence of the TSR2. Perhaps there is even repression of "the threat" in the present case.

27 This is like the self-defeating effects of safety systems as described by Charles Perrow (1984). For further commentary, see Law 2000b.

28 The battle takes place on a narrow terrain: on whether or not to add "the social" to "the technical."

29 Ludmilla Jordanova (1980, 45), discussing the similarities of science and medicine in the eighteenth century, writes that these were "activities . . . associated with sexual metaphors which were clearly expressed in designating nature as a woman to be unveiled, unclothed, and penetrated by masculine science." In the present context, the metaphor applies to the military conquest of the enemy. I have cited the case of "penetration." The reconnaissance function of the aircraft might be treated as a form of "unclothing."

30 Or, alternatively, one could say childbirth, which would certainly destabilize the distribution away from the impersonal but would also turn the aircraft into a gendered female entity. It is then possible to tell quite pointed stories about male birthing. See Easlea 1983. I explore this possibility briefly later.

31 Humans are often said to be language users, or they are said to have intentions, or again, they are said to have emotions or souls. For discussion of the first two, see Callon and Law 1995.

32 "Fantasy can be satisfied by realist means which secure the credibility of the imaginary scene, with details of costume, setting, accessories. The realist mode of signification disguises the fact of production beneath its veneer of appearance" (Pollock 1988, 24). The argument is like that of Roland

Barthes in *Myth Today* (1973), where he explores the way in which the picture of a black soldier saluting the Tricolor naturalizes mythological work.

33 Perhaps, then, this is a *punctum* in the sense intended by Roland Barthes? "The second element will break (or punctuate) the *studium*. This time it is not I who seek it out (as I invest the field of my *studium* with my sovereign consciousness), it is this element which rises from the scene, shoots out of it like an arrow, and pierces me" (Barthes 1993, 26). Punctum, interpellation.

34 I am grateful to Mark Elam, who is exploring the connections between the iconography of military operations and medicine and who sensitized me to the possibilities of such a link.

35 Consider the analysis of pre-Raphaelite painting in Pollock 1988.

36 There are further complexities. I am grateful to Bernike Pasveer for pointing out that it is a man who enters the "cockpit," who masters the femininity of the aircraft. But this suggests that the aircraft achieves its power (as opposed to its potency) only when the male is dominant.

37 The term comes from Donna Haraway (1991b) and has been developed by Marilyn Strathern (1991).

38 Much writing in art theory has shown that the differences are complex and context dependent. But in the present context I would start (as I did in chapter 2) by pointing to the initially linear complexities of the textual strategies of containment, and contrast these with the ways in which the visual depictions can perform several contrasts simultaneously. It must not, however, be supposed that narrative is necessarily linear and visual depiction nonlinear.

7. Decisions

1 This is the place where actor-network theory tends to distinguish itself from (other versions of?) poststructuralism: its concern with the character of translation (or distribution) as this moves through a range of materials quite removed from language. For commentary, see Latour 1993 and Law 1994.

2 The major difference being that this time there is no mention of buying British aircraft.

3 Although I am building an argument here that draws on certain semiotic assumptions, its form is not so far from the interpretive or *verstehende* argument developed by Barry Barnes (1988), who treats society as a substantially shared distribution of knowledge.

4 The term "scenario" is used by Michel Callon in his work on the *véhicule éléctrique* (1980).

5 It is standard that there are techniques in Politics for not taking decisions, that is, not taking decisions yet. This form of deferral, which displaces de-

cidability into the future, is being performed in this case. The possibility of taking an option on the American F111A aircraft means that there is no need to decide whether or not to buy it yet. See exhibit 7.5 for the way this works. And note that it constitutes current *un*decidability within a framework of future *decidability*.

6 In which case it is possible to imagine the decision and its context as a monad. For this argument in the context of organizational learning, see Law 2000a.

7 Helen Verran, describing Australian aboriginal cosmologies, redescribes what are usually called "dreamings" as "epistemic imaginaries." Her point is, in part, that aborigines do not distinguish in the clear way being described here between that which is real and that which is imaginary. For them it is indeed possible to perform "imaginaries" and so make them real, and not to perform realities and so lose them. The point is that aboriginal cosmology takes reality and performance as going together. They cannot be pried apart. The argument is developed in a broader context in Watson 1989, Watson-Verran 1995, and Verran 1998. For commentary, see Law and Benschop 1997.

8 The importance of the distinction between regions, networks, and fluids is explored in Mol and Law 1994.

9 Perhaps, however, it is one of the divisions performed in Euro-American culture, though many seek to break it. I think, for instance, of science fiction writers, and the way they have been explored by such commentators as Donna Haraway (1997).

10 The implication is that Britain might no longer undertake military commitments in the Far East.

11 Similar, if not identical, concerns are addressed by the notion of "heterogeneous engineering" in which different kinds of material considerations are drawn together and managed. But, as I tried to show in chapter 5, it becomes important to be heterogeneous in the ways in which we imagine heterogeneity. See Lyotard 1991.

12 The point is recognized but addressed in somewhat different terms in political theory, for instance as the "mobilization of bias" that keeps certain items off the political agenda. Or as the inability of agents to voice—or even recognize—their "real interests." For discussion that covers these two possibilities while favoring the latter, see Lukes 1974.

13 Homage, here, to Michel Foucault (1979) who made speakable this reality, which was always a reality, of the "microphysics of power." And to those many writers and practitioners, Marxist and feminist, poststructuralist and otherwise, who have sought to find ways of telling of the political character of everyday practice.

14 Though having noted that the class of aircraft are made to be passive, we need also to note that there are (as always) ambivalences. For instance Crossman sometimes complains that the decisions are being made else-

where (exhibits 7.9 and 7.10), so converting himself into a passive agent. Or at any rate, into one who cannot act as he would like. As background for exhibit 7.9, note that George Wigg was paymaster general, essentially the minister for the army. And Chequers is the country house available for use by the British prime minister.

15 Michel Callon (1986b) explores the way in which agents constitute themselves by performing strategies that tend, recursively, to turn them into obligatory points of passage.

16 Remember, obligatory points of passage are achievements and not naturally occurring and stable features. And "power" does not exist, of itself, but rather it is created or borrowed. The point is eloquently developed by Bruno Latour in his *Irréductions* (1988a).

17 This is the term used by Barry Barnes in his discussion of powers (1988). But there are similarities between his position and that adopted by Callon (1991). See also Law 1991a.

18 See, for instance, Miller 1992. Note that there is a huge industry seeking to reproduce the reasoning of government. Perhaps its efforts fall into three broad camps. First and simply, to understand why governments made the decisions that they did. The exhibits are drawn from this large literature, though in the present context one would want to add several more important sources, including Hastings 1966 and Williams, Gregory, and Simpson 1969. Second, much of that literature also has a normative content. For if the government reasons in a discretionary manner and it does so in a way that is controversial within the framework of High Politics, then there is an important space for remedial reasoning: exploring why it was that the government got it wrong. Such explanations often work in terms of a depth hermeneutics of (more or less concealed) social interests of the kind described by Barnes (1977) and discussed in a different context in chapter 4. This, for instance, is one of the explanatory modes adopted by the Conservative MP Stephen Hastings, and is no doubt an expression of what I characterized as "policy narrative" in chapter 3. Third, there is a class of literature that, while also normative, seeks to understand why or how the government got itself into the position where it was obliged to make a controversial and unsatisfactory decision. This literature removes itself in some degree from political partisanship and explores the workings of the government machine. Or, to put it in the language I'm using here, considers the workings of that machine as a center of translation. The monograph by Williams, Gregory, and Simpson falls predominantly into this category. Though there is much to be learned from each of these approaches and perhaps especially from the third, which in some measure considers what I am here calling the "political," they all work on, and so reproduce, the assumption that Politics is properly constituted as centered decision making about important matters.

19 I might mention, in addition, the great wall of official secrecy. In a study

such as this, which deals with confidential matters, much of the material that one would like to see is either not publicly available, or, if it is available, can only be seen through unofficial channels. The existence of this great wall leaves the investigator with two choices. To seek official approval for the investigation and whatever is written or to avoid doing so. Both have effects that tend to reproduce the distribution between what is confidential and what is public. Note that one of the features of academic writing, the convention of "properly" citing sources, merely sharpens the division between public and confidential. Journalists who have no such requirement are frequently able to use confidential material in open publication without difficulty, in part because the confidentiality of its origin is not acknowledged. Which is not to say that dissident journalists necessarily find life any easier than dissident academics. Life can be very hard for both.

20 The strategic distinction between ends and means. For instance, it is remarkably difficult to persuade nonpractitioners that "technology" is an interesting object in its own right rather than a more or less (un)satisfactory means to the important end of social life.

21 See Mol 1994 and 1999; Mol and Berg 1994; Berg and Mol 1998; and Mol and Law 1994.

22 This point is established in the network theory of representation developed by such writers as Mary Hesse (1974) and also implicit in the overall approach to singularity and multiplicity developed in this book.

23 The analysis also works for the aircraft themselves. Are we dealing with one TSR2 or many? As I have argued, the answer is both—or an oscillation between the two. Michel Callon and I have explored this point in a somewhat similar way elsewhere (Law and Callon 1989), talking of an aircraft of variable geometry that changed its shape as time went by. Here, however, I am suggesting that the aircraft has *simultaneous* multiple geometries. Though if the argument about the simultaneous performance of complexity and simplicity is sustained, no doubt it was also being performed simply, as possessing a single geometry, for instance, as an aircraft that was altogether too expensive. This then would be a fractional aircraft.

24 The metaphor of play is taken from Jean-François Lyotard 1991. And the notion of fluid is developed in an analogous topological mode in Mol and Law 1994.

25 See Cussins 1998a; Haraway 1991b, 1991c, 1991d; Latour 1993, 1996a; Mol 1008b, 2002a; Berg and Mol 1998; Singleton 1996, 1998; Law and Singleton 2000a; Dugdale 1999; Moser 2000; Watson-Verran and Turnbull 1995; Turnbull 2000, and Verran 1998.

8. Arborescences

1 What follows is retold in various forms in many locations. My own efforts include Law 1988, and Law and Callon 1988, 1989, and 1992. Others in-

clude Hastings 1966, Williams, Gregory, and Simpson 1969, Gardner 1981, and Wood 1975 and 1986.

2 More systematic or authorized versions of these exchanges can be found in Gardner 1981 and Williams, Gregory, and Simpson 1969.

3 Bruno Latour perhaps comes close to something like this in his *Aramis* (1996a).

4 The term "detail," which I have already used in my interpolated comments on the basic TSR2 story, also performs its own distribution between big and small, which correlates with important and unimportant—the performance of the hierarchy that is a characteristic of treelike structures. This point is developed in earlier chapters, by Law (2000c), and in a different context by Strathern (1991).

5 The importance of this point was brought home to me in discussion with Annemarie Mol and Bob Cooper. See Law 1994, 84 and chapter 7 (this volume).

6 There are many sociologies that tell this story in one way or another: symbolic interaction, game-theoretical models, the theories of power developed by Barry Barnes. These tell the story that in the social, whatever is real is whatever is *believed* to be real. Part of the trick is going to be to extend this recognition into the performances and narrations made in different materialities. For the construction of a sociology that assumes the social world is a shared distribution of knowledge, see Barnes 1988.

7 The actor-network approach is explored at some length by Bruno Latour (1987). For recent discussion see the collected papers in Law and Hassard 1999. For Donna Haraway, see the papers collected in Haraway 1991c and 1997.

8 For a beautifully subtle discussion of the complexities of insider/outsider distinctions, see Mol 1994.

9 The references are to fractal mathematics with its capacity to generate self-similar patterns at every level of scale. For popular accounts, see Gleick 1987 and Kellert 1993.

9. Pinboards

1 I thank Celia Lury for helping me to think about the logic of pinboards. For details, see Lury 1998.

2 See Jean-François Lyotard 1984b, Mike Featherstone 1991, Jean Baudrillard 1988a, Mark Poster 1990, and Sherry Turkle 1996.

3 See, for instance, the writing of Anthony Giddens 1990, Ulrich Beck 1992, and Beck, Giddens, and Scott Lash 1994.

4 Exceptions include Haraway (1997), Hetherington (1997c, 1999), Hetherington and Lee (2000), Latour (1996a), Law (1994, 1998a), Mol (2002a), Strathern (1991), and Taussig (1993), and the poststructuralist philosophers, including perhaps particularly Derrida (1978).

5 As I indicated at the beginning of the book, I am, of course, aware that the word *singularity* implies specificity, even uniqueness. Throughout, however, I have sought to bend the term by setting it in tension with the many-ness of *plurality*. Used in the way I want to deploy it, singularity thus refers to ontological continuity or conformable and uniquely accountable conditions of narration.

6 See, for instance, Annemarie Mol 2002a, Marianne de Laet and Annemarie Mol 2000, and Tiago Moreira 1998.

Primary Sources

In the course of the research for this book, primary sources were explored at a number of locations. Those directly referred to in this book are located in three archives:

—The Public Record Office near Kew in West London. This is the source of all public documents.

—The Historical Retrieval Information Centre, Public Relations Department, British Aerospace, Warton Aerodrome, Warton, near Preston, Lancashire. This is the source of all English Electric documents.

—The Brooklands Museum in Weybridge in Surrey, which is where records from British Aerospace plc, Weybridge, were transferred. This is the source of all Vickers Armstrong documents.

Public Record Office, United Kingdom

AIR8/2002. 1954. Air Ministry internal document, from CAS to DLD, 22 January 1954.

AIR8/2014. 1955. *Future Development Programme.* Letter from G. W. Tuttle, D.C.A.S., to Secretary of State, 19 October.

AIR8/2014. 1956. *Future Tactical Air Bombardment/Reconnaissance Policy.* Document drafted by Operational Requirement Branch, Air Staff, 19 October.

AIR8/2167. 1957. Letter from George Ward, Minister for Air, to Duncan Sandys, Secretary of State for Defence, 26 February.

AIR8/2196. 1958a. *OR 339 Development and Production.* Note attached to note from G. W. Tuttle, D.C.A.S., to Secretary of State, 18 December.

AIR8/2196. 1958b. *Historic Diary Tactical Strike Reconnaissance Aircraft.* Undated note attached to letter from G. W. Tuttle, D.C.A.S., to C.A.S., 1 August.

Air Ministry 1958. *Air Staff Requirement No. O.R.339, Tactical Strike/Reconnaissance Aircraft.* London: Air Ministry.

——. 1959. *Air Staff Requirement No. O.R./343.* London: Air Ministry.

Historical Retrieval Information Centre, Public Relations Department, British Aerospace, Warton Aerodrome, Warton, near Preston, Lancashire

English Electric. 1957. *Note on Crew Fatigue due to Low Altitude Flying for G.O.R. 339.* Warton: English Electric.

——. 1959. *Project 17.* Warton: English Electric.

English Electric/Short Bros. 1958. *Aircraft Project P.17; Tactical Strike/Reconnaissance Aircraft to G.O.R. 339.* 3 vols. Mimeo.

Brooklands Museum at Weybridge, Surrey

British Aircraft Corporation. 1962. *TSR2 Weapons System.* Weybridge: British Aircraft Corporation.

Vickers Armstrong. 1958a. *G.O.R. 339: One Engine versus Two Engines: a Proposal for a Small Working Party to Study Available Statistical Records.* Memorandum from Mr. J. K. Quill to Mr. H. Gardner, 5 May.

———. 1958b. *Study of G.O.R. 339.* Weybridge: Vickers Armstrong (Aircraft) Ltd.

———. 1958c. *G.O.R. 339.* Letter from J. K. Quill to H. H. Gardner, 1 July.

———. 1958d. *G.O.R. 339 Weapons System.* Short brochure for the Type 571, Weybridge: Vickers Armstrong, n.d., but probably early 1958.

———. 1958e. *G.O.R. 339: Single versus Twin Engines: Bombing/Navigation Accuracy.* Memorandum, 29 May.

Secondary Sources

Abercrombie, Nicholas, Stephen Hill, and Bryan S. Turner. 1980. *The Dominant Ideology Thesis.* London: George Allen and Unwin.

Alpers, Svetlana. 1988. *Rembrandt's Enterprise: The Studio and the Market.* Chicago: Chicago University Press.

———. 1989. *The Art of Describing: Dutch Art in the Seventeenth Century.* London: Penguin.

Althusser, Louis. 1971. "Ideology and Ideological State Apparatuses (Notes towards an Investigation)," pp. 121–73 in *Lenin and Philosophy and Other Essays.* Ed. L. Althusser. London: New Left Books.

Anderson, Benedict. 1991. *Imagined Communities: Reflections on the Origin and Spread of Nationalism.* 2d ed. London: Verso.

Ashmore, Malcolm. 1989. *The Reflexive Thesis: Wrighting Sociology of Scientific Knowledge.* Chicago: Chicago University Press.

———. 1993. "Behaviour Modification of a Catflap: A Contribution to the Sociology of Things." *Kennis en Methode* 17:214–29.

Ashmore, Malcolm, Michael J. Mulkay, and Trevor J. Pinch. 1989. *Health and Efficiency: A Sociology of Health Care Economics.* Milton Keynes, UK: Open University Press.

Auster, Paul. 1988. *The Invention of Solitude.* London: Faber and Faber.

Austin, J. L. 1965. *How to Do Things with Words.* Ed. J. O. Urmston. Oxford: Oxford University Press.

———. 1970. *Philosophical Papers.* Ed. J. O. Urmston and G. J. Warnock. Oxford: Oxford University Press.

Bal, Mieke. 1991. *Reading Rembrandt: Beyond the Word-Image Opposition.* Cambridge: Cambridge University Press.

Barnes, Barry. 1977. *Interests and the Growth of Knowledge.* London: Routledge and Kegan Paul.

——. 1988. *The Nature of Power*. Cambridge, UK: Polity Press.

Barthes, Roland. 1973. "Myth Today," pp. 109–59 in *Mythologies*. Trans. Annette Lavers. St. Albans: Paladin.

——. 1977. "From Work to Text," pp. 155–64 in *Image, Music, Text*. Trans. Stephen Heath. St. Albans: Paladin.

. 1993. *Camera Lucida: Reflections on Photography*. London: Vintage.

Baudrillard, Jean. 1988. "Simulacra and Simulations," pp. 166–84 in *Selected Writings*. Cambridge, UK: Polity Press.

Bauman, Zygmunt. 1989. *Modernity and the Holocaust*. Cambridge, UK: Polity Press.

Beamont, Roland. 1968. *Phoenix into Ashes*. London: William Kimber.

Beck, Ulrich. 1992. *Risk Society: Towards a New Modernity*. Trans. Mark Ritter. London: Sage.

Beck, Ulrich, Anthony Giddens, and Scott Lash. 1994. *Reflexive Modernisation: Politics, Tradition and Aesthetics in the Modern Social Order*. Cambridge, UK: Polity Press.

Benjamin, Walter. 1985. *One Way Street and Other Writings*. London: Verso.

——. 1992. *Illuminations*. Trans. Harry Zohn. London: Fontana.

Berg, Marc, and Annemarie Mol, eds. 1998. *Differences in Medicine: Unraveling Practices, Techniques, and Bodies*. Durham, N.C.: Duke University Press.

Bijker, Wiebe E. 1987. "The Social Construction of Bakelite: Towards a Theory of Invention," pp. 159–87 in *The Social Construction of Technical Systems: New Directions in the Sociology and History of Technology*. Ed. Wiebe E. Bijker, Thomas P. Hughes, and Trevor Pinch. Cambridge: MIT Press.

——. 1992. "The Social Construction of Fluorescent Lighting, or How an Artefact Was Invented in Its Diffusion Stage," pp. 75–102 in *Shaping Technology, Building Society: Studies in Sociotechnical Change*. Ed. Wiebe E. Bijker and John Law. Cambridge: MIT Press.

Bourdieu, Pierre. 1984. *Distinction: a Social Critique of the Judgement of Taste*. London: Routledge.

Braidotti, Rosi. 1994. *Nomadic Subjects: Embodiment and Sexual Difference in Contemporary Feminist Theory*. Gender and Culture. New York: Columbia University Press.

Brenna, Brita, John Law, and Ingunn Moser, eds. 1998. *Machines, Agency and Desire*. TMV Report Series. Oslo: University of Oslo.

Bryson, Norman. 1990. *Looking at the Overlooked: Four Essays on Still Life Painting*. London: Reaktion Books.

Buck-Morss, Susan. 1989. *The Dialectics of Seeing: Walter Benjamin and the Arcades Project*. Cambridge: MIT Press.

Callon, Michel. 1980. "Struggles and Negotiations to Define What Is Problematic and What Is Not: The Sociology of Translation," pp. 197–219 in *The Social Process of Scientific Investigation: Sociology of the Sciences Year-*

book, vol. 4. Ed. Karin D. Knorr, Roger Krohn, and Richard D. Whitley. Dordrecht: Reidel.

———. 1986a. "The Sociology of an Actor-Network: The Case of the Electric Vehicle," pp. 19–34 in *Mapping the Dynamics of Science and Technology: Sociology of Science in the Real World.* Ed. Michel Callon, John Law, and Arie Rip. London: Macmillan.

———. 1986b. "Some Elements of a Sociology of Translation: Domestication of the Scallops and the Fishermen of Saint Brieuc Bay," pp. 196–233 in *Power, Action, and Belief: A New Sociology of Knowledge?* Sociological Review Monograph, no. 32. Ed. John Law. London: Routledge and Kegan Paul.

———. 1991. "Techno-economic Networks and Irreversibility," pp. 132–61 in *A Sociology of Monsters? Essays on Power, Technology, and Domination.* Sociological Review Monograph, vol. 38. Ed. John Law. London: Routledge.

Callon, Michel, and Bruno Latour. 1981. "Unscrewing the Big Leviathan: How Actors Macrostructure Reality and How Sociologists Help Them to Do So," pp. 277–303 in *Advances in Social Theory and Methodology: Toward an Integration of Micro- and Macro-Sociologies.* Ed. Karin D. Knorr-Cetina and Aaron V. Cicourel. Boston: Routledge and Kegan Paul.

Callon, Michel, and John Law. 1995. "Agency and the Hybrid Collectif." *South Atlantic Quarterly* 94:481–507.

Cohn, Carol. 1987. "Sex and Death in the Rational World of Defense Intellectuals." *Signs: Journal of Women in Culture and Society* 12:687–718.

———. 1993. "Wars, Wimps and Women: Talking Gender and Thinking War," pp. 227–46 in *Gendering War Talk.* Ed. Miriam Cooke and Angela Woollacott. Princeton: Princeton University Press.

Cooke, Miriam, and Angela Woollacott, eds. 1993. *Gendering War Talk.* Princeton, N.J.: Princeton University Press.

Cooper, Robert. 1986. "Organization/Disorganization." *Social Science Information* 25:299–335.

———. 1989a. "Modernism, Post Modernism and Organizational Analysis 3: The Contribution of Jacques Derrida." *Organization Studies* 10:479–502.

———. 1989b. "The Visibility of Social Systems," pp. 51–59 in *Operational Research in the Social Sciences.* Ed. M. C. Jackson, P. Keys, and S. Cropper. New York: Plenum.

Cooper, Robert, and John Law. 1995. "Organization: Distal and Proximal Views," pp. 275–301 in *Research in the Sociology of Organizations.* Studies of Organizations in the European Tradition, vol. 13. Ed. Samuel B. Bachrach, Pasquale Gagliardi, and Bryan Mundell. Greenwich, Conn.: JAI Press.

Crossman, Richard. 1975. *Diary of a Cabinet Minister.* Vol. 1, *Minister of Housing, 1964–1966.* London: Hamish Hamilton and Jonathan Cape.

Cussins, Adrian, et al. 1992. *Registration Marks: Metaphors for Subobjectivity.* London: Pomeroy Purdy Gallery.

Cussins, Charis M. 1998a. "Ontological Choreography: Agency for Women Patients in an Infertility Clinic," pp. 166–201 in *Differences in Medicine: Unraveling Practices, Techniques, and Bodies.* Ed. Marc Berg and Annemarie Mol. Durham, N.C.: Duke University Press.

———. 1998b. "'Quit Sniveling, Cryo-Baby. We'll Work Out Which One's Your Mama!,'" pp. 40–66 in *Cyborg Babies: From Techno-Sex to Techno-Tots.* Ed. Robbie Davis-Floyd and Joseph Dumit. New York: Routledge.

Dawson, Graham. 1994. *Soldier Heroes: British Adventure, Empire and the Imagining of Masculinities.* London: Routledge.

Defence. 1965. "Statement by the Ministry of Defence on the Cancellation of the TSR.2—6th April, 1965." London: Ministry of Defence.

de Laet, Marianne, and Annemarie Mol. 2000. "The Zimbabwe Bush Pump: Mechanics of a Fluid Technology." *Social Studies of Science* 30:225–63.

de Lauretis, Teresa. 1987. *Technologies of Gender: Essays on Theory, Film, and Fiction.* London: Macmillan.

Deleuze, Gilles, and Félix Guattari. 1988. *A Thousand Plateaus: Capitalism and Schizophrenia.* London: Athlone.

Deleuze, Gilles, and Claire Parnet. 1987. *Dialogues.* London: Athlone.

der Derian, James. 1992. *Antidiplomacy: Spies, Terror, Speed, and War.* Cambridge, Mass.: Blackwell.

Derrida, Jacques. 1978. *Writing and Difference.* London: Routledge.

Douglas, Mary. 1982. "Cultural Bias," pp. 183–254 in *In the Active Voice.* London: Routledge and Kegan Paul.

Dugdale, Anni. 1999. "Materiality: Juggling Sameness and Difference," pp. 113–35 in *Actor Network Theory and After.* Ed. John Law and John Hassard. Oxford: Blackwell and the Sociological Review.

Dunning, William W. 1991. *Changing Images of Pictorial Space: A History of Spatial Illusion in Painting.* Syracuse, N.Y.: Syracuse University Press.

Easlea, Brian. 1983. *Fathering the Unthinkable: Masculinity, Scientists, and the Nuclear Arms Race.* London: Pluto.

Edgerton, Samuel Y. 1976. *The Renaissance Rediscovery of Linear Perspective.* New York: Harper and Row.

Elam, Mark. 1999. "Living Dangerously with Bruno Latour in a Hybrid World." *Theory, Culture and Society* 16:1–29.

Elias, Norbert. 1978. *The History of Manners.* Oxford: Blackwell.

———. 1983. *The Court Society.* Oxford: Basil Blackwell.

Featherstone, Mike. 1991. *Consumer Culture and Postmodernism.* London: Sage.

Fisher, Philip. 1991. *Making and Effacing Art: Modern American Art in a Culture of Museums.* New York: Oxford University Press.

Forty, Adrian. 1986. *Objects of Desire: Design and Society, 1750–1980.* London: Thames and Hudson.

Foucault, Michel. 1970. *The Order of Things: An Archaeology of the Human Sciences.* London: Tavistock.

———. 1972. *The Archaeology of Knowledge.* London: Tavistock.

———. 1976. *The Birth of the Clinic: An Archaeology of Medical Perception.* London: Tavistock.

———. 1979. *Discipline and Punish: The Birth of the Prison.* Harmondsworth, UK: Penguin.

———. 1981. *The History of Sexuality,* Volume 1: *An Introduction.* Harmondsworth, UK: Penguin.

———. 1986. "Of Other Spaces." *Diacritics* 16:22–27.

Gardner, Charles. 1981. *British Aircraft Corporation: A History.* London: Batsford.

Garfinkel, Harold. 1967. *Studies in Ethnomethodology.* Englewood Cliffs, N.J.: Prentice Hall.

Geertz, Clifford. 1988. *Works and Lives: The Anthropologist as Author.* Cambridge, UK: Polity.

Giddens, Anthony. 1990. *The Consequences of Modernity.* Cambridge, UK: Polity Press.

Gleick, James. 1987. *Chaos: Making a New Science.* London: Sphere Books.

Gombrich, E. H. 1989. *The Story of Art.* 15th ed. London: Phaidon.

Gusterson, Hugh. 1995a. "Becoming a Weapons Scientist," pp. 255–74 in *Technoscientific Imaginaries: Conversations, Profiles, and Memoirs.* Ed. George E. Marcus. Chicago: University of Chicago Press.

———. 1995b. "Short Circuit: Watching Television with a Nuclear Weapons Scientist," pp. 107–17 in *The Cyborg Handbook.* Ed. Chris Hables Gray. New York: Routledge.

———. 1998. *Nuclear Rites: A Weapons Laboratory at the End of the Cold War.* Berkeley: University of California Press.

Hansard (1964–65). *Hansard* 710 (1964–65): col. 1258, April 13, 1965.

Haraway, Donna. 1989. *Primate Visions: Gender, Race, and Nature in the World of Modern Science.* London: Routledge and Chapman Hall.

———. 1991a. "The Biopolitics of Postmodern Bodies: Constitutions of Self in Immune System Discourse," pp. 203–30 in *Simians, Cyborgs, and Women: The Reinvention of Nature.* London: Free Association Books.

———. 1991b. "A Cyborg Manifesto: Science, Technology, and Socialist Feminism in the Late Twentieth Century," pp. 149–81 in *Simians, Cyborgs, and Women: The Reinvention of Nature.* London: Free Association Books.

———. 1991c. *Simians, Cyborgs, and Women: The Reinvention of Nature.* London: Free Association Books.

———. 1991d. "Situated Knowledges: The Science Question in Feminism and the Privilege of Partial Perspective," pp. 183–201 in *Simians, Cyborgs, and Women: The Reinvention of Nature.* London: Free Association Books.

———. 1992. "The Promises of Monsters: A Regenerative Politics for Inappropriate/d Others," pp. 295–337 in *Cultural Studies*. Ed. Lawrence Grossberg, Cary Nelson, and Paul Treichler. New York: Routledge.

———. 1994. "A Game of Cats Cradle: Science Studies, Feminist Theory, Cultural Studies." *Configurations* 1:59–71.

———. 1997. *Modest_Witness@Second_Millennium.Female_Man©_Meets Oncomouse™: Feminism and Technoscience*. New York: Routledge.

Hastings, Stephen. 1966. *The Murder of TSR-2*. London: Macdonald.

Hennion, Antoine. 1996. "Les Jambes d'Hercule: Des Oeuvres et du Gout," pp. 309–21 in *Représenter, Hybrider, Coordoner*. Ed. Cécile Méadel and Vololona Rabeharisoa. Paris: École des Mines de Paris.

Hesse, Mary B. 1974. *The Structure of Scientific Inference*. London: Macmillan.

Hetherington, Kevin. 1997a. *The Badlands of Modernity: Heterotopia and Social Ordering*. London: Routledge.

———. 1997b. "In Place of Geometry: the Materiality of Place," pp. 183–99 in *Ideas of Difference: Social Spaces and the Labour of Division*. Sociological Review Monograph. Ed. Kevin Hetherington and Rolland Munro. Oxford: Blackwell.

———. 1997c. "Museum Topology and the Will to Connect." *Journal of Material Culture* 2:199–218.

———. 1998. *Expressions of Identity: Space, Performance, Politics*. London: Sage.

———. 1999. "From Blindness to Blindness: Museums, Heterogeneity, and the Subject," pp. 51–73 in *Actor Network Theory and After*. Ed. John Law and John Hassard. Oxford and Keele: Blackwell and the Sociological Review.

Hetherington, Kevin, and John Law. 1998. "Allegory and Interference: Representation in Sociology," http://www.comp.lancs.ac.uk/sociology/reskhjl1.html.

Hetherington, Kevin, and Nick Lee. 2000. "Social Order and the Blank Figure." *Society and Space* 18:169–84.

Hirschauer, Stefan, and Annemarie Mol. 1995. "Shifting Sexes, Moving Stories: Feminist/Constructivist Dialogues." *Science, Technology and Human Values* 20:368–85.

HMSO. 1955. "The Supply of Military Aircraft." *Cmd. 9388*. London: Her Majesty's Stationery Office.

Hughes, Thomas P. 1979. "The Electrification of America: The System Builders." *Technology and Culture* 20:124–61.

———. 1983. *Networks of Power: Electrification in Western Society, 1880–1930*. Baltimore, Md.: Johns Hopkins University Press.

Jameson, Fredric. 1991. *Postmodernism, or, The Cultural Logic of Late Capitalism*. Durham, N.C.: Duke University Press.

Jay, Martin. 1993. *Downcast Eyes: The Denigration of Vision in Twentieth-Century French Thought*. Berkeley: University of California Press.

Jordanova, Ludmilla J. 1980. "Natural Facts: A Historical Perspective on Science and Sexuality," pp. 42–69 in *Nature, Culture, and Gender.* Ed. Carol MacCormack and Marilyn Strathern. Cambridge: Cambridge University Press.

Kellert, Stephen H. 1993. *In the Wake of Chaos.* Chicago: University of Chicago Press.

Knorr-Cetina, Karin D. 1992. "The Couch, the Cathedral and the Laboratory: On the Relationship between Experiment and Laboratory in Science," pp. 113–38 in *Science as Practice and Culture.* Ed. Andrew Pickering. Chicago: University of Chicago Press.

——. 1996. "The Care of the Self and Blind Variation: An Ethnography of the Empirical in Two Sciences," pp. 287–310 in *The Disunity of Science Boundaries, Contexts, and Power.* Ed. Peter Galison and D. Stamp. Stanford, Calif.: Stanford University Press.

Kuhn, Thomas S. 1970. *The Structure of Scientific Revolutions.* Chicago: University of Chicago Press.

Latour, Bruno. 1987. *Science in Action: How to Follow Scientists and Engineers through Society.* Milton Keynes, UK: Open University Press.

——. 1988a. "Irréductions," pp. 151–236 in *The Pasteurization of France.* Cambridge: Harvard University Press.

——. 1988b. *The Pasteurization of France.* Cambridge: Harvard University Press.

——. 1988c. "The Politics of Explanation: An Alternative," pp. 155–76 in *Knowledge and Reflexivity: New Frontiers in the Sociology of Knowledge.* Ed. Steve Woolgar. London: Sage.

——. 1990. "Drawing Things Together," pp. 19–68 in *Representation in Scientific Practice.* Ed. Michael Lynch and Steve Woolgar. Cambridge: MIT Press.

——. 1993. *We Have Never Been Modern.* Brighton, UK: Harvester Wheatsheaf.

——. 1996a. *Aramis, or the Love of Technology.* Cambridge: MIT Press.

——. 1996b. *Petite Réflexion sur le Culte Moderne des Dieux Faitiches.* Paris: Les Empêcheurs de Penser en Rond.

——. 1999a. "On Recalling ANT," pp. 15–25 in *Actor Network and After.* Ed. John Law and John Hassard. Oxford: Blackwell and the Sociological Review.

——. 1999b. *Politiques de la Nature: Comment Faire Entrer les Sciences en Démocratie.* Paris: La Découverte.

Law, John. 1987. "Technology and Heterogeneous Engineering: the Case of the Portuguese Expansion," pp. 111–34 in *The Social Construction of Technical Systems: New Directions in the Sociology and History of Technology.* Ed. Wiebe E. Bijker, Thomas P. Hughes, and Trevor Pinch. Cambridge: MIT Press.

——. 1988. "The Anatomy of a Sociotechnical Struggle: The Design of the

TSR2," pp. 44–69 in *Technology and Social Process.* Ed. Brian Elliot. Edinburgh: Edinburgh University Press.

——. 1991a. "Introduction: Monsters, Machines, and Sociotechnical Relations," pp. 1–23 in *A Sociology of Monsters: Essays on Power, Technology, and Domination.* Sociological Review Monograph, vol. 38. Ed. John Law. London: Routledge.

Law, John, ed. 1991b. *A Sociology of Monsters: Essays on Power, Technology, and Domination.* Sociological Review Monograph, vol. 38. London: Routledge.

Law, John. 1992. "The Olympus 320 Engine: A Case Study in Design, Development, and Organisational Control." *Technology and Culture* 33:409–40.

——. 1994. *Organizing Modernity.* Oxford: Blackwell.

——. 1996a. *Air Show.* Keele, UK: Keele University.

——. 1996b. "Organizing Accountabilities: Ontology and the Mode of Accounting," pp. 283–306 in *Accountability: Power, Ethos, and the Technologies of Managing.* Ed. Jan Mouritsen and Rolland Munro. London: International Thompson Business Press.

——. 1997. "Traduction/Trahison: Notes on ANT." TMV Working Paper 106. Oslo: University of Oslo. http://www.comp.lancs.ac.uk/sociology/stslaw2.html.

——. 1998a. "After Metanarrative: On Knowing in Tension," pp. 88–108 in *Into the Realm of Organisation: Essays for Robert Cooper.* Ed. Robert Chia. London: Routledge.

——. 1998b. "Machinic Pleasures and Interpellations," pp. 23–48 in *Machines, Agency, and Desire.* Ed. Brita Brenna, John Law, and Ingunn Moser. Oslo: University of Oslo.

——. 1999. "After ANT: Topology, Naming and Complexity," pp. 1–14 in *Actor Network Theory and After.* Ed. John Law and John Hassard. Oxford: Blackwell and the Sociological Review.

——. 2000a. "Comment on Suchman, and Gherardi and Nicolini: Knowing as Displacing." *Organization* 7:349–54.

——. 2000b. "Ladbroke Grove, or How to Think about Failing Systems." http://www.comp.lancs.ac.uk/sociology/soc055jl.html.

——. 2000c. "Transitivities." *Society and Space* 18:133–48.

Law, John, and Ruth Benschop. 1997. "Resisting Pictures: Representation, Distribution and Ontological Politics," pp. 158–82 in *Ideas of Difference: Social Spaces and the Labour of Division.* Sociological Review Monograph. Ed. Kevin Hetherington and Rolland Munro. Oxford: Blackwell.

Law, John, and Michel Callon. 1988. "Engineering and Sociology in a Military Aircraft Project: A Network Analysis of Technical Change." *Social Problems* 35:284–97.

——. 1989. "On the Construction of Sociotechnical Networks: Content and Context Revisited." *Knowledge and Society* 9:57–83.

——. 1992. "The Life and Death of an Aircraft: A Network Analysis of Technical Change," pp. 21–52 in *Shaping Technology, Building Society: Studies in Sociotechnical Change*. Ed. Wiebe Bijker and John Law. Cambridge: MIT Press.

Law, John, and John Hassard, eds. 1999. *Actor Network Theory and After*. Oxford and Keele: Blackwell and the Sociological Review.

Law, John, and Annemarie Mol. 1998. "On Metrics and Fluids: Notes on Otherness," pp. 20–38 in *Organised Worlds: Explorations in Technology, Organisation, and Modernity*. Ed. Robert Chia. London: Routledge.

Law, John, and Ingunn Moser. 1999. "Managing, Subjectivities, and Desires." *Concepts and Transformation* 4:249–79.

Law, John, and Vicky Singleton. 2000a. "Allegory and Its Others." http://www.comp.lancs.ac.uk/sociology/soc051jl.html.

———. 2000b. "Performing Technology's Stories." *Technology and Culture* 40: 765–75.

———. 2000c. "This Is Not an Object." http://www.comp.lancs.ac.uk/sociology/soc032jl.html.

Lee, Nick, and Steve Brown. 1994. "Otherness and the Actor Network: The Undiscovered Continent." *American Behavioral Scientist* 36:772–90.

Lively, Penelope. 1991. *City of the Mind*. London: Penguin.

Lukes, Steven. 1974. *Power: A Radical View*. London: Macmillan.

Lury, Celia. 1998. *Prosthetic Culture: Photography, Memory, and Identity*. London: Routledge.

Lyotard, Jean-François. 1984a. *Driftworks*. Foreign Agents Series. New York: Semiotext(e).

———. 1984b. *The Postmodern Condition: A Report on Knowledge*. Manchester, UK: Manchester University Press.

———. 1985. *Discours, Figure*. Paris: Editions Klincksieck.

———. 1991. *The Inhuman: Reflections on Time*. Cambridge, UK: Polity Press.

Lyotard, Jean-François, and Jean-Loup Thébaut. 1985. *Just Gaming*. Manchester, UK: University of Manchester.

MacCormack, Carol, and Marilyn Strathern, eds. 1980. *Nature, Culture and Gender*. Cambridge: Cambridge University Press.

MacKenzie, Donald. 1990. *Inventing Accuracy: A Historical Sociology of Nuclear Missile Guidance*. Cambridge: MIT Press.

MacKenzie, Donald, and Judy Wajcman, eds. 1985. *The Social Shaping of Technology: How the Refrigerator Got Its Hum*. Milton Keynes: Open University Press.

———. 1999. *The Social Shaping of Technology: How the Refrigerator Got Its Hum*. Rev. ed. Milton Keynes, UK: Open University Press.

Mannheim, Karl. 1953. "Conservative Thought," pp. 74–164 in *Essays on Sociology and Social Psychology*. Ed. Karl Mannheim. London: Routledge and Kegan Paul.

Martin, Emily. 1994. *Flexible Bodies*. Boston: Beacon Press.

Miller, Peter. 1992. "Accounting and Objectivity: The Invention of Calculating Selves and Calculable Spaces." *Annals of Scholarship* 9:61–86.

Mitchell, W. J. T. 1994. *Picture Theory.* Chicago: University of Chicago Press.

Mol, Annemarie. 1994. "Where the Strangers Are." Mimeo. University of Limburg.

——. 1998a. "Lived Reality and the Multiplicity of Norms: A Critical Tribute to George Canguilhem." *Economy and Society* 27:274–84.

——. 1998b. "Missing Links, Making Links: The Performance of Some Artheroscleroses," pp. 144–65 in *Differences in Medicine: Unraveling Practices, Techniques, and Bodies.* Ed. Annemarie Mol and Marc Berg. Durham, N.C.: Duke University Press.

——. 1999. "Ontological Politics: A Word and Some Questions," pp. 74–89 in *Actor Network Theory and After.* Ed. John Law and John Hassard. Oxford: Blackwell and the Sociological Review.

——. 2002a. *The Body Multiple: Ontology in Medical Practice.* Durham, N.C.: Duke University Press.

——. 2002b. "Cutting Surgeons, Walking Patients: Some Complexities Involved in Comparing," in *Complexities in Science, Technology, and Medicine.* Ed. John Law and Annemarie Mol. Durham, N.C.: Duke University Press.

Mol, Annemarie, and Marc Berg. 1994. "Principles and Practices of Medicine: The Coexistence of Various Anaemias." *Culture, Medicine and Psychiatry* 18:247–65.

Mol, Annemarie, and Bernard Elsman. 1996. "Detecting Disease and Designing Treatment: Duplex and the Diagnosis of Diseased Leg Vessels." *Sociology of Health and Illness* 18:609–31.

Mol, Annemarie, and John Law. 1994. "Regions, Networks, and Fluids: Anaemia and Social Topology." *Social Studies of Science* 24:641–71.

——. 2001. "Situated Bodies and Distributed Selves: Enacting Hypoglycaemia." http://www.comp.lans.ac.uk/sociology/stslaw5.html.

Mol, Annemarie, and Jessica Mesman. 1996. "Neonatal Food and the Politics of Theory: Some Questions of Method." *Social Studies of Science* 26:419–44.

Moreira, Tiago. 1988. "How Many Bodies Are There?" Paper delivered at the Wetenschap, Technologie en Moderne Cultuur/Centre de Sociologie de l'Innovation Conference on the Body, École des Mines, Paris, June 1998.

Moser, Ingunn. 1998. "Kyborgens Rehabilitering," pp. 39–74 in *Betatt av viten. Bruksanvisninger til Donna Haraway.* Ed. Kristin Asdal, Anne-Jorunn Berg, Brita Brenna, Ingunn Moser, and Linda Rustad. Oslo: Spartacus.

——. 2000. "Against Normalisation: Subverting Norms of Ability and Disability." *Science as Culture* 9:201–40.

Moser, Ingunn, and John Law. 1998. "Materiality, Textuality, Subjectivity: Notes on Desire, Complexity, and Inclusion." *Concepts and Transforma-*

tion: International Journal of Action Research and Organizational Renewal 3:207–27.

———. 1999. "Good Passages, Bad Passages," pp. 196–219 in *Actor Network and After.* Ed. John Law and John Hassard. Oxford and Keele: Blackwell and the Sociological Review.

Mouritsen, Jan, and Rolland Munro, eds. 1996. *Accountability: Power, Ethos, and the Technologies of Managing.* London: International Thompson Business Press.

Nathan, Tobie, and Isabelle Stengers. 1995. *Médecins et Sorciers.* Les Plessis-Robinson: Les Empêcheurs de Penser en Rond.

Perrow, Charles. 1984. *Normal Accidents: Living with High Risk Technologies.* New York: Basic Books.

Pickering, Andrew. 1993. "The Mangle of Practice: Agency and Emergence in the Sociology of Science." *American Journal of Sociology* 99:559–89.

———. 1995. *The Mangle of Practice: Time, Agency, and Science.* Chicago: University of Chicago Press.

Pollock, Griselda. 1988. *Vision and Difference: Femininity, Feminism, and the Histories of Art.* London: Routledge.

Poster, Mark. 1990. *The Mode of Information: Poststructuralism and Social Context.* Cambridge, UK: Polity Press.

Rabinow, Paul. 1989. *French Modern: Norms and Forms of the Social Environment.* Cambridge: MIT Press.

Ransom, Stephen, and Robert Fairclough. 1987. *English Electric Aircraft and Their Predecessors.* London: Putnam.

Reed, Bruce, and Geoffrey Williams. 1971. *Denis Healey and the Policies of Power.* London: Sidgewick and Jackson.

Robins, Ken, and Les Levidow. 1995. "Socializing the Cyborg Self: The Gulf War and Beyond," pp. 119–25 in *The Cyborg Handbook.* Ed. Chris Hables Gray. New York: Routledge.

Rosenberg, Stanley D. 1993. "The Threshold of Thrill: Life Stories in the Skies over Southeast Asia," pp. 43–66 in *Gendering War Talk.* Ed. Miriam Cooke and Angela Woollacott. Princeton, N.J.: Princeton University Press.

Rotman, Brian. 1987. *Signifying Nothing: The Semiotics of Zero.* Stanford, Calif.: Stanford University Press.

Said, Edward W. 1991. *Orientalism: Western Conceptions of the Orient.* London: Penguin.

Sapolsky, Harvey. 1972. *The Polaris System Development: Bureaucratic and Programmatic Success in Government.* Cambridge: MIT Press.

Serres, Michel. 1974. *La Traduction, Hermes III.* Paris: Les Éditions de Minuit.

———. 1992. *Michel Serres: Entretiens avec Bruno Latour.* Paris: François Bourin.

Shapin, Steven. 1994. *A Social History of Truth: Civility and Science in Seventeenth-Century England.* Chicago: Chicago University Press.

Shapin, Steven, and Simon Schaffer. 1985. *Leviathan and the Air Pump:*

Hobbes, Boyle, and the Experimental Life. Princeton, N.J.: Princeton University Press.

Singleton, Vicky. 1993. "Science, Women and Ambivalence: An Actor-Network Analysis of the Cervical Screening Campaign." Ph.D. diss. University of Lancaster.

———. 1996. "Feminism, Sociology of Scientific Knowledge and Postmodernism: Politics, Theory and Me." *Social Studies of Science* 26:445–68.

———. 1998. "Stabilizing Instabilities: The Role of the Laboratory in the United Kingdom Cervical Screening Programme," pp. 86–104 in *Differences in Medicine: Unraveling Practices, Techniques, and Bodies.* Ed. Marc Berg and Annemarie Mol. Durham, N.C.: Duke University Press.

Singleton, Vicky, and Mike Michael. 1993. "Actor-networks and Ambivalence: General Practitioners in the UK Cervical Screening Programme." *Social Studies of Science* 23:227–64.

Staudenmaier, John, SJ. 1994. "To Fall in Love with the World: Individualism and Self Transcendence in American Life." *Studies in the Spirituality of Jesuits* 26:1–28.

Stone, Allucquère Rosanne. 1991. "Will the Real Body Please Stand Up? Boundary Stories about Virtual Cultures," pp. 81–118 in *Cyberspace: First Steps.* Ed. Michael Benedikt. Cambridge: MIT Press.

———. 1995a. "Innocence and Awakening: Cyberdämmerung at the Achibe Research Laboratory," pp. 177–95 in *Technoscientific Imaginaries: Conversations, Profiles, and Memoirs.* Ed. George E. Marcus. Chicago: University of Chicago Press.

———. 1995b. "Split Subjects, Not Atoms; or, How I Fell in Love with My Prosthesis," pp. 393–406 in *The Cyborg Handbook.* Ed. Chris Hables Gray. New York: Routledge.

———. 1995c. *The War of Desire and Technology at the Close of the Mechanical Age.* Cambridge: MIT Press.

Strathern, Marilyn. 1991. *Partial Connections.* Savage, Md.: Rowman and Littlefield.

———. 1992. *After Nature: English Kinship in the Late Twentieth Century.* Cambridge: Cambridge University Press.

———. 1996. "Cutting the Network." *Journal of the Royal Anthropological Institute* 2:517–35.

———. 1999. "What Is Intellectual Property After?" pp. 156–80 in *Actor Network and After.* Ed. John Law and John Hassard. Oxford: Blackwell and the Sociological Review.

Taussig, Michael. 1993. *Mimesis and Alterity: A Particular History of the Senses.* New York: Routledge.

Theweleit, Klaus. 1987. *Male Fantasies.* 2 vols. Cambridge, UK: Polity Press.

Traweek, Sharon. 1988a. *Beamtimes and Lifetimes: The World of High Energy Physics.* Cambridge: Harvard University Press.

———. 1988b. "Feminist Perspectives on Science Studies." *Science, Technology and Human Values* 13:250–53.

———.1992. "Border Crossings: Narrative Strategies in Science Studies and among Physicists in Tsukuba Science City, Japan," pp. 429–66 in *Science as Practice and Culture.* Ed. Andrew Pickering. Chicago: University of Chicago Press.

———. 1995a. "Bachigai (Out of Place) in Ibaraki: Tsukuba Science City, Japan," pp. 355–77 in *Technoscientific Imaginaries: Conversations, Profiles, and Memoirs.* Ed. George E. Marcus. Chicago: University of Chicago Press.

———. 1995b. "Bodies of Evidence: Law and Order, Sexy Machines, and the Erotics of Fieldwork among Physicists," pp. 211–25 in *Choreographing History.* Ed. Susan Leigh Foster. Bloomington: Indiana University Press.

———. 1999. "Pilgrim's Progress: Male Tales Told During a Life in Physics," pp. 525–42 in *The Science Studies Reader.* Ed. Mario Biagioli. New York: Routledge.

Turkle, Sherry. 1984. *The Second Self: Computers and the Human Spirit.* New York: Simon and Schuster.

———. 1996. *Life on the Screen: Identity in the Age of the Internet.* London: Weidenfeld and Nicolson.

Turnbull, David. 1993a. "The Ad Hoc Collective Work of Building Gothic Cathedrals with Templates, String, and Geometry." *Science, Technology and Human Values* 18:315–40.

———. 1993b. *Maps Are Territories, Science Is an Atlas.* Chicago: University of Chicago Press.

———. 1996. "Cartography and Science in Early Modern Europe: Mapping the Construction of Knowledge Spaces." *Imago Mundi* 48:5–24.

———. 2000. *Masons, Tricksters and Cartographers: Comparative Studies in the Sociology of Scientific and Indigenous Knowledge.* Amsterdam: Harwood Academic Publishers.

Verran, Helen. 1998. "Re-Imagining Land Ownership in Australia." *Postcolonial Studies* 1:237–54.

Wagner, Roy. 1991. "The Fractal Person," pp. 159–73 in *Big Men and Great Men: Personifications of Power in Melanesia.* Ed. Maurice Godelier and Marilyn Strathern. Cambridge: Cambridge University Press.

Wajcman, Judy. 1991. *Feminism Confronts Technology.* Cambridge, UK: Polity Press.

———. 1998. *Managing Like a Man: Women and Men in Corporate Management.* Cambridge, UK: Polity Press.

Watson, Helen, the Yolngu Community, and David Wade Chambers. 1989. *Singing the Land, Signing the Land.* Geelong, Victoria: Deakin University Press.

Watson-Verran, Helen, and David Turnbull. 1995. "Science and Other Indigenous Knowledge Systems," pp. 115–39 in *Handbook of Science and Tech-*

nology Studies. Ed. Sheila Jasanoff, Gerard E. Markle, James C. Petersen, and Trevor Pinch. Thousand Oaks, Calif.: Sage.

Weber, Max. 1978. *Economy and Society: An Outline of Interpretive Sociology.* Ed. Guenther Roth and Claus Wittich. Berkeley: University of California Press.

Wheelock, Arthur K., ed. 1995. *Johannes Vermeer.* Washington, D.C.: National Gallery of Art.

Williams, Geoffrey, Frank Gregory, and John Simpson. 1969. *Crisis in Procurement: A Case Study of the TSR 2.* London: Royal United Service Institution.

Wilson, Harold. 1971. *The Labour Government, 1964–1970: A Personal Record.* London: Weidenfeld and Nicolson and Michael Joseph.

Wood, Derek. 1986. *Project Cancelled.* 2d ed. London: Janes.

Woolgar, Steve, ed. 1988a. *Knowledge and Reflexivity: New Frontiers in the Sociology of Knowledge.* London: Sage.

Woolgar, Steve. 1988b. *Science: The Very Idea.* Chichester: Ellis Horwood and Tavistock.

——. 1991. "Configuring the User: The Case of Usability Trials," pp. 58–99 in *A Sociology of Monsters? Essays on Power, Technology, and Domination.* Sociological Review Monograph, vol. 38. Ed. John Law. London: Routledge.

Zuckerman, Solly. 1988. *Monkeys, Men, and Missiles: An Autobiography 1946–88.* London: Collins.

Abercrombie, Nick, 199–200
Absence, 92, 94–96, 101–103, 110, 113–114, 123, 160, 195; fear, 101–102; material, 97–98; Otherness, 102, 195; of Russians, 101. *See also* Heterogeneity; Oscillation: between absence and presence; Presence
Accountability, 150–151, 181
Actor-network theory, 54, 58, 60, 66, 84, 95, 98, 112, 180, 206 n.10, 212 nn.22–23, 213 n.4, 218 n.1, 222 n.7; black boxing in, 95, 184; *interessement* in, 84; obligatory passage point in, 152, 220 nn.15–16; punctualization in, 95; scenario in, 217 n.24, 218 n.4; translation in, 98, 213 n.5, 218 n.1. *See also* Immutable mobiles; Technology studies
Aerodynamic center. *See* Aerodynamics: aerodynamic center in
Aerodynamics, 91–111: aerodynamic center in, 108–109; altitude in, 100; angle of attack in, 93–94; aspect ratio in, 100; gust response in, 92–111, 129; lift slope in, 92–94, 106–110; planform in, 107–110; size and, 103–106; speed in, 99–103
Aesthetics, 31, 64, 115–142, 150; agency and, 121–125; as "detail," 120; function versus, 118, 120, 215 n.8; of juxtaposition, 189; narrative form and, 58–59, 61; obviousness and, 64; technics and, 116–120, 131, 134, 140, 142, 156, 214 n.3, 215 nn.6 and 12
Agency: active versus passive, 121–122, 125, 127, 216 n.17, 219 n.14; aesthetics and, 121–125; constituted, 220 n.15; decision making and, 151–153; distributed, 122, 127, 132–133, 216 n.17; female, 138–139; heroic male, 31, 123–125, 132–133, 208 nn.8 and 31; movement and, 121–122; nature

and, 136; technical, 125–127; TSR2 as having, 122, 129, 133, 137, 216 n.17. *See also* Gender; Subjectivity
Aircraft. *See* Canberra; F111; F111A; F111K; Hunter; Meteor; NA 39; P1; P.17A; Phantom; Spitfire; Type 571: design of; Valiant; V bombers; Viking; Viscount
Aircraft industry, 171–172; rationalization 172, 178–180, 183
Airfields, 27–28, 100, 135–137
Air Ministry, 105–106
Allegory, 4, 5, 39, 191 n.7, 208 n.1
Alpers, Svetlana, 23–24, 42–43, 134–135, 216 n.19
Alterity, 114, 115. *See also* Otherness
Althusser, Louis, 38, 50–51, 52–53, 62, 88, 216 n.20
Altitude, 30; low, 31, 67, 100
Ambivalence, 114, 162, 219 n.14
Angle of attack. *See* Aerodynamics: angle of attack in
Anodyne narrative, 132
ANT. *See* Actor-network theory
Aramis, 222 n.3
Arborescence, 5, 6, 10–11, 55, 163–164, 173–175, 180, 188, 206 n.5, 222 n.4; hierarchy and, 20, 184; project-ness and, 183–185
Archaeology: Foucault and, 49
Arteriosclerosis, 16–17, 208 n.18
Art history. *See* History of art
Ashmore, Malcolm, 190–191
Aspect ratio. *See* Aerodynamics: aspect ratio in
Auster, Paul, 126
Austin, J. L., 11, 176

Bailly, David, 42–43, 58
Bal, Mieke, 134–135
Barnes, Barry, 218 n.3, 220 nn.17–18, 222 n.6
Barthes, Roland, 217 n.32, 218 n.33
Bauman, Zygmunt, 1, 90, 200, 216 n.22
Beamont, Roland, 58, 210 n.28

Benjamin, Walter, 208 nn.1 and 13
Bias, cultural. *See* Cultural bias
Bias, mobilization of, 219 n.12
Binarisms and interdependence:
absence/presence, 92, 94–96,
101–103, 110, 113–114, 123–
124, 160, 195; active/passive,
121–122, 125, 127, 129, 216 n.17;
aesthetics/function, 118, 120,
215 n.8; aesthetics/technics,
116–120, 131, 134, 140, 142, 156,
214 n.4, 215 nn.6 and 12; cul-
ture/nature, 125; discourse/figure,
189; dynamics/statics, 121–
122, 136; fantasy/reality, 10,
147–148; female/male, 31, 123–
125, 132–133, 208 n.8, 210
n.31; human/nonhuman, 81,
90, 133, 151, 162, 213 n.2, 217
n.31; impurity/purity, 142; in-
sider/outsider, 222 n.8; invul-
nerable/vulnerable, 124–125,
129–131; multiplicity/singularity,
3, 53, 86, 111, 115, 126, 142, 159–
160, 163, 175, 183, 187, 193,
202; nature/technics, 131; ob-
ject/subject, 43, 45, 133, 174–175;
pictures/text, 10, 117–118, 121,
150, 196, 215 n.15, 218 n.38;
private/public, 38, 42–44, 58,
154–155, 220 n.19
Black boxing. *See* Actor-network
theory: black boxing in
Blackburn Aircraft, 211 n.8
Body, the, 141, 212 n.21; absent,
98; Althusser and, 50; Foucault
on, 48–49; method and, 62; sub-
ject/object distinction and, 43
Braque, Georges, 117–118, 196, 215
n.15
Bristol Aircraft, 211 n.9
Bristol Siddeley, 74. *See also* Olympus
British Aircraft Corporation, 12,
116–117, 172, 182
British government: ballistic missile
policy, 83; cancellation decision,
143–162; industrial merger policy,

80, 172, 178–180, 183; procure-
ment policy, 66–67, 75, 164–173;
size of aircraft, 82; weapons
system policy, 75, 111
Brochure. *See* P1; TSR2; Type 571:
design of
Brown, George, 144
Brush fire wars, 169
Buccaneer. *See* NA 39
Bureaucratic politics, 84, 171–172

Cabinet. *See* TSR2: cancellation of
Calculation. *See* Narrative
Callaghan, James, 144
Callon, Michel, 57–58, 220 n.15, 221
n.23
Campaign for Nuclear Disarma-
ment, 40
Canberra, 66–69, 71, 101, 165, 169
Canberra replacement aircraft, 110,
169, 178. *See also* P.17A; TSR2;
Type 571: design of
Cancellation of TSR2. *See* TSR2:
cancellation of
Cartography, 23–25, 206 n.7; subjec-
tivity and, 23–25
Cathedrals, 209 n.11
Centering, 112–113, 115, 153, 159,
193
Cervical Screening Programme, 207
n.16, 208 n.19
Cézanne, Paul, 118
Childbirth, 217 n.30
Chronology. *See* Genealogy; Narra-
tive: chronology and; Project-ness;
Time
Civilizing process, 43–48
Coherence. *See* Coordination;
Fractionality: and coherence;
Noncoherence; Oscillation
Collusion, 9; narrative and political,
7–8, 10–11, 164, 185
Colonialism, 202
Communion as dream, 99
Complexity: interference and, 135,
140–142; pinboard and, 198–199;
as self-limiting, 95
Conditions of possibility, 8, 9, 53,

153, 163, 175, 183; performed, 148; singular, 183

Confession. *See* Personal, the

Configuring the user, 212 n.20

Constatives, 176–178, 180

Consumption versus production, 134

Continuity. *See* Cultural bias; Narrative: continuity in; Project-ness: continuity and

Cooper, Bob, 222 n.5

Coordination, 143

Coordination, strategies of, 15–35, 53, 92; cartographic, 23–25; perspectival, 21–23; physical structure, 18; pictorial, 29–32; syntactical, 18; system, 26–28; tabular, 18–21

Corporeality. *See* Body, the

Cost-effect analysis. *See* Design, aircraft: cost-effect analysis

Coyote, 162

Criticism: analytical, 7; modernity and, 86; political, 7; postmodernism and, 87–88

Crossman, Richard, 144–145, 149, 152–155, 157, 160, 219 n.14

Culture, 70, 79, 125; as similarity and difference, 76

Cultural bias, 113, 119, 175, 193, 212 n.14; to continuity, 9, 70, 76, 79, 84–87, 175, 185; and project-ness, 87, 175; to smoothness, 119, 175, 185, 189, 195

Cussins, Charis, 162

Cybernetics, 28, 207 n.10

Cyborg, 3, 162

Dalyell, Tam, 57

Dawson, Graham, 210 n.30

De Haviland, 211 n.9

Decision making, 10, 143–162, 174, 186, 201, 219 n.6; agency created in, 151–153; accountable similarity in, 150; centering and, 181–182; "detail" created in, 153–156; deferral, 218 n.5; deletions in, 150; difference effaced in, 156–160; fantasy and, 10, 147–148;

interpellation and, 51; options created in, 149–151; as oscillation between multiplicity and singularity, 159–160, 163; perspectivalism and, 158–159; places of, 151–153; public/private division and, 154–155; quantified, 183; rational, 181, 183; reality and, 10, 147–148. *See also* TSR2: cancellation of

Deferral, 20, 90, 92, 114, 131; heterogeneity and, 110; material, 206 n.6; "Politics" as, 218 n.5; research and development as, 132; technics as, 131, 147. *See also* Différance

Deletion, 150; of difference, 156–160; of "politics," 148–151

Deleuze, Gilles, 1, 64, 163, 173, 175

Depth hermeneutics. *See* Narrative: depth hermeneutics and

Derrida, Jacques, 20, 115

Design, aircraft, 66–84; cost-effect analysis, 73, 76–82; cultures, 77; engine number, 80–83, 104–106, 214 n.9; heterogeneity, 91–92; lethality and aircraft size, 76–82, 105–106, 113; modernism and, 113, 212 n.17; size, 76–77, 79–84, 103–106; vulnerability, 76–77; wing, 91–111. *See also* Aerodynamics; Heterogeneity; P.17A; TSR2; Type 571: design of; Weapons system

"Detail," 10, 20, 117–118, 122, 150, 153–156, 163, 184, 215 n.6, 216 n.17, 222 n.4; aesthetics as, 120; as distribution, 118; enacted in decision making, 153–156. *See also* Scale

Deterrence. *See* Independent nuclear deterrent

Development batch, 168, 172

Différance, 115. *See also* Deferral

Difference, 70, 97, 161; in culture, 76; effaced, 156–160; problem of, 15, 15–17, 38, 65, 192. *See also* Multiplicity

Diffraction. *See* Interference

Discontinuity. *See* Narrative: discontinuity

Discourse: and figure, 189–190. *See also* Episteme; Foucault, Michel; Narrative forms

Distribution, 61–62, 88, 120, 133, 140, 143, 158, 206 n.12; of agency, 122, 127, 216 n.17; into "details," 118; design and, 94; enacted, 65, 118; between fantasy and reality, 10, 147–148; as fractional effect, 200; gender and, 141; heterogeneity and, 96; places of power, 151–153; political, 199–201; between politics and the nonpolitical, 156; between public and secret, 220 n.19. *See also* Binarisms and interdependence

Domesticity, 138

Dominant ideology thesis, 199–200

Dominant narrative thesis, 200–201

Douglas, Mary, 212 n.14

Drawing things together, 205 n.6

Dualisms. *See* Binarisms and interdependence

Dugdale, Anni, 162

Edwards, Sir George, 80

Elam, Mark, 218 n.34

Elias, Norbert, 46–48, 49, 61

Empirical philosophy, 126

Empiricism, 210 nn.25–26

Engine number. *See* Design: engine number

English Electric, 66–84, 91–111, 113, 211 n.3, 213 n.3; merger with Vickers Armstrong, 80

Enlightenment, 45, 127, 131, 190, 216 n.22

Episteme, 52–53, 62, 161, 209 n.20, 213 nn.28 and 30

Epistemic imaginaries, 162, 219 n.7

Epistemology, 8, 36–37, 63, 65, 158–159, 177, 180, 182, 185–186; feminist, 188. *See also* Ontology

Erotics. *See* Subjectivity: erotic

Esoteric narrative, 57–58, 59–60, 70, 210 n.26

Essentialism, 199

Established disorder, 3, 39, 50, 126, 203

Ethics: as narrative form, 56–57, 59–60

Ethnicity, 200

Ethnography: personal in, 38–39, 42; vanity, 43, 48, 208 n.9

Ethnomethodology, 206 n.2

F111, 146

F111A, 145, 146, 149, 151, 157–158, 173; cancellation of order, 173

F111K, 181

Fantasy, 10, 163; as opposed to reality, 147–148, performed, 147–148

Feminist epistemology. *See* Epistemology: feminist

Feminist theory, 205 n.8, 213 n.1

Figure and discourse, 189–190

Fisher, Philip, 216 n.18

Flâneurie, 189–190

Fluidities, 203, 219 n.8, 221 n.24

Forty, Adrian, 119

Foucault, Michel, 18, 49–50, 52–53, 61–62, 160–161, 163, 198, 209 n.20, 219 n.13; and discourses, 48–49; and microphysics of power, 160–161, 163, 210 n.10

Fractals, 205 n.4, 222 n.9

Fractionality 3, 4, 11, 88, 114, 115, 160–162, 182, 196–197, 200, 203; and coherence, 2, 4–5, 8; enacted, 34; as more than one and less than many, 3, 4; politics and, 8; quietism and, 11; rhizome and, 5; writing and, 5–6. *See also* Oscillation

Freud, Sigmund, 47

Function: versus aesthetics, 118, 120, 215 n.6

Fyfe, Gordon, 47

Game theory, 222 n.5

Gardner, Charles, 73, 179, 182–183

Gender, 10, 31, 40–41, 52, 138–140,

141, 196, 199, 208 n.8, 209 n.19,
210 n.31, 215 nn.11 and 13, 216
n.18, 217 nn.26 and 30, 218 n.36;
and aesthetic narrative, 58, 214
n.3; dangerous female sexuality
and potency of TSR2, 138–140;
science and, 217 n.29. *See also*
Male subjectivities: heroic; Patriarchy; Sexuality
Genealogy: narrative and, 9, 71–73,
85–86, 134, 185
General Operational Requirement,
213 n.6
Gibbs, Martin, 209 n.17
Giddens, Anthony, 90
God-eye. *See* God-trick
God-trick, 41–44, 48, 89, 185, 210
n.25
GOR 339, 55, 67–69, 71–73, 82, 100–
101, 104, 110, 166, 168–172, 178,
180; submissions, 172
Grand narrative, 1, 86, 87, 90, 189,
213 n.30; and politics, 200; of
TSR2 project, 164–173
Gregory, Frank, 33–34, 180–181, 220
n.18
Guattari, Félix, 1, 173, 175
Gulf War, 122, 173
Gust response. *See* Aerodynamics:
gust response in

Haraway, Donna, 3, 38, 41–42, 46,
59, 126, 162, 180, 185, 188–189,
203, 212 n.14, 219 n.9
Hastings, Stephen, 56, 74, 220 n.18
Haviland, Denis, 179
Hawker Group, 211 n.9
Healey, Denis, 144–146, 155, 157
Hennion, Antoine, 186
Hermeneutics. *See* Narrative: depth
hermeneutics and
Heroism. *See* Male subjectivities:
heroic
Hesse, Mary, 221 n.22
Heterogeneity, 195, 214 n.14, 219
n.11; absence/presence, 95–96,
101–103, 110–112, 123, 160; de-
ferral, 110; design, 91, 92; hetero-

geneity/materiality, 97–98, 112;
heterogeneity/noncoherence,
105–106; heterogeneity/simplicity,
95–96; 97–98, 112; modernism
and, 113–114; ontological, 112;
Otherness and, 101–102. *See also*
Heterogeneous engineering
Heterogeneous engineering, 44, 54,
112–113, 219 n.11; as centering,
113; as oscillation, 113
Heterotopia, 53, 62, 210 n.21
Hetherington, Kevin, 208 n.1, 210
n.21
Hierarchy: arborescence and, 184;
tables and, 20
Hill, Stephen, 199–200
Hirschauer, Stefan, 141
History, plain: as narrative form, 55–
56, 59–60, 69–70, 173, 183, 210
n.26
History of art, 118, 134, 215 n.10, 217
n.26, 218 n.35
History of TSR2. *See* TSR2: his-
tory of
Humans: and nonhumans, 81, 90,
133, 151, 162, 213 n.2, 217 n.31;
dissolved in modernism, 113
Hunter, Hawker, 167, 211 n.13
Husbandry, 138

Ideological state apparatuses, 50, 52,
216 n.20
Ideology, 50, 52–53, 62; dominant
ideology thesis, 199–200
Illustrations. *See* Visual materials
Immutable mobiles, 28
Impartiality, 214 n.7
Impurity: versus purity, 142
Independent nuclear deterrent, 40,
164–166
Indigenous knowledge traditions,
162
Indirection, 38–39, 191
Interessement. See Actor-network
theory: *interessement* in
Interests: and narrative, 85–86;
social, 69, 84–86, 134, 158, 185,
220 n.18

Interference, 7, 53, 92, 115, 125, 126, 135, 143, 152, 156, 160, 180, 182, 196, 214 n.2; narrative, 5, 59–61, 176, 178, 210 n.29; obdurate, 202; the personal and, 63; productive, 185; strength of, 135, 140–142; subjectivities and, 59–64

Interpellation, 9, 49–51, 53, 55–58; narrative, 55–61; and obviousness, 52, 64; and technology, 52, 64

Intertextuality. *See* Narrative: intertextuality

Interviewees, 54

Invulnerability: versus vulnerability, 123–125, 129

Irréductions, 126, 202, 216 n.20, 220 n.16

Johns, Jasper, 206 n.6, 216 n.18

Jenkins, Roy, 144

Jesse, tree of, 72, 174

Jordanova, Ludmilla, 217 n.29

Juxtaposition. *See* Aesthetics: of juxtaposition; Pinboard: as juxtaposition

Knorr-Cetina, Karin, 27

Knowing, 186, 190–193, 195, 197; Foucault's analysis, 49; foundationalism and, 5; fractional coherence and, 4–5; reality as made in, 63; tension in, 191

Knowledge. *See* Knowing

Korean War, 167–168

Kuhn, Thomas, 216 n.22

Laboratory witnessing, 208 n.7

Large technical systems, 26, 212 n.22

Latour, Bruno, 28, 90, 98, 126, 162, 175, 202, 207 n.11, 213 n.2, 216 n.20, 220 n.16, 222 n.3

Law, John, 57–58, 112, 186, 221 n.23

Lethality. *See* Design, aircraft: lethality and aircraft size

Lift slope. *See* Aerodynamics: lift slope in

Lightning. *See* P1

Linear perspective. *See* Perspectivalism

Little narratives, 164; TSR2 and, 178–180

Lively, Penelope, 89

London Transport, 119

Lowe, Adam, 126

Lukes, Steven, 219 n.12

Lury, Celia, 222 n.1

Lyotard, Jean-François, 1, 3, 173, 190, 199, 221 n.24

Male subjectivities: heroic, 29–32, 40–41, 52, 122–125, 132–133, 138, 196, 208 n.8, 210 n.31, 215 n.14. *See also* Agency: heroic male; Subjectivity: male

Maps. *See* Cartography

Materiality, 47–49, 62, 66, 206 n.6, 208 n.11, 219 n.10; heterogeneity and, 97–98, 112

McNeil, Maureen, 210 n.27, 214 n.3

Meteor, 98

Method, 161; the body and, 62; impartiality, 214 n.7; and interviewees, 54; obviousness and, 64; pinboard as, 197; secrecy and, 220 n.19; symmetry in, 127–128. *See also* Naive reader: as method

Microphysics of power, 160, 163, 219 n.13

Miller, Peter, 220 n.18

Ministry of Aviation, 170

Ministry of Defence, 145

Ministry of Supply, 55, 170, 179

Missiles: ballistic, 83, 166, 168, 172; guidance systems, 212 n.25; surface to air, 67, 130, 165, 215 n.12; V2, 166

Mitchell, J. W. T., 115

Mode of ordering, 216 n.22

Modern project. *See* Modernity: modern project; Project-ness: modernism and

Modernism, 1, 2, 3, 5, 6, 11, 90–91, 113, 115, 119, 142, 164, 195, 198, 212 n.19; heterogeneity and, 113–114; purification and, 115,

118, 215 n.7; style and, 119; utopianism and, 113–114; weapons systems and, 78–79, 212 n.17. *See also* Postmodernism

Modernity, 53, 198, 200, 213 n.2; critical, 86; modern project, 87, 89, 114, 143

Monadology, 213 n.4, 219 n.6

Mol, Annemarie, 15–16, 65, 126, 141, 161, 186, 192, 208 n.18, 210 n.32, 222 n.5

Moser, Ingunn, 141, 162

Mulkay, Michael, 190–191

Multiplicity, 2, 8, 11, 32–33, 38, 53, 59–61, 65, 86, 88, 90, 115, 126, 142, 143, 159–161, 183, 198; decisions as, 159–160; effaced, 90; marginalized, 76, 114; politics and, 161–162. *See also* Coordination; Difference; Interference; Obduracy: multiplicity and; Objects: as multiple; Oscillation: between singularity and multiplicity; Singularity; Singularity and multiplicity; Subjectivity: as multiple

Mundanity, 137, 207 n.13; versus transcendence, 123–124

Munro, Rolland, 211 n.10

Musgrave, Sir Cyril, 178–180, 182–183

NA 39, 82–83, 170–173

Naive reader, 13–15; as method, 15, 32

Narrative: aesthetic, 58–59, 61; anodyne, 132; arborescent, 173–174, 180, 188; calculable, 79, 150, 183, 185; causal, 70, 174; chronology and, 9, 55, 71–72, 174, 185–187; coherence in, 159; collusion in, 7–8, 10–11, 164, 185; complexity in, 4, 5; continuity in, 9, 70–71, 74, 76, 79, 85, 185, 188; cultural bias in, 76, 84–86; depth hermeneutics and, 9, 84–86, 118, 220 n.18; discontinuity in, 74–75; dominant narrative thesis,

200–201; esoteric, 57–60; ethical, 56–57, 59–60; forms, 55–58; genealogy as, 9, 71–73, 85–86; homogenization, 79; illustrations and, 10; indirection and, 38–39; interference and, 5, 59–61, 178, 180, 210 n.29; interpellations, 55–56; intertextuality, 180, 210 n.25; normative, 56; origin stories as, 70, 73, 85, 211 n.7; as performing reality, 6, 7, 10–11, 38–39, 52, 65, 144, 174–182; plain history, 55–56, 59–60, 69–70, 173, 183, 210 n.26; policy, 56, 58–60, 68, 70, 220 n.18; politics and, 7–8; potential, 206 n.6; project-ness, 87; redemption, 189, 193, 195; representing reality, 6; as rhizome, 5, 6; scale and, 9; singularity, 76, 157–159, 190; smoothness, 119, 175, 185, 189, 193, 195; social interests, 84–85, 158; synchronous, 78, 84; system, 85–86; system as, 9, 78–79; technology studies and, 70, 84; totalizing, 85. *See also* Anodyne narrative; Genealogy; Grand narrative; Little narrative; Origin stories

Narrative forms, 55–58; aesthetic, 58–59, 61; esoteric, 57–60; ethical, 56–57, 59–60; plain history, 55–56, 59–60, 69–70, 173, 183, 210 n.26; policy, 56, 58–60, 68, 70, 220 n.18

Nathan, Tobie, 143

Nature, 125; made active, 128, 136; as passive/vulnerable, 140; as a resource, 127

Navigators, 97

Noncoherence, 125, 159, 175; as heterogeneity, 105–106; of objects, 33–35; strength of, 140–141, 201–202

Nonhumans. *See* Humans

Nonmodernism, 90–91, 191, 210 n.22

Normativity. *See* Narrative: normative

Nuclear deterrent. *See* Independent nuclear deterrent; V bombers

Nuclear weapons, 7, 13, 34, 67, 164–166

Obduracy: multiplicity and, 11; reality and, 6. *See also* Obdurate incoherence

Obdurate incoherence, 200–202

Objectivity: as situated, 41, 44

Objects: as agents, 122, 127, 129; as centered, 21; as decentered, 3, 195; as fractional, 34–35, 160; as fractionally coherent, 3, 4, 11, 35, 193, 194, 196–197; as multiple, 2, 32, 143; as noncoherent, 33–35; and singularity, 3, 33–35, 164; and subjects, 43, 45, 133, 174; virtual, 186. *See also* TSR2

Obligatory passage point. *See* Actor-network theory: obligatory passage point in

Obviousness. *See* Interpellation

Official Secrets Act, 192, 220 n.19

Olympus engine, 74

Ontological politics, 198

Ontology, 5, 8, 36–37, 63, 65, 89, 159, 162, 177, 182, 185–186, 198; heterogeneity, 112; performed, 65, 147–148. *See also* Epistemology

Operational Requirement, 213 n.6. *See also* GOR 339

Operational Requirements Branch. *See* Royal Air Force: Operational Requirements Branch

Optics, 41

OR 339. *See* GOR 339

Origin stories, 70, 73, 85, 211 n.7. *See also* Narrative: origin stories

Oscillation, 89, 114, 126, 159–160, 197, 214 n.2; between absence and presence, 9–10, 92, 94–96, 98, 101–103, 110–114; decision making and, 159–160; distribution and, 96; heterogeneous engineering and, 113; between modernism and postmodernism, 90; between picture and text, 10, 29, 116; between singularity and multiplicity, 10, 90–91, 142, 163, 175, 187, 193, 202

Otherness, 101–103, 123, 159–160, 195, 214 n.8; tellability, 102; threat, 102. *See also* Alterity

Out-thereness, 24, 177. *See also* Realism

P1, 67–69, 81

P1.B. *See* P1

P.17A, 67–71, 74, 79–82, 211 n.6; brochure, 67–71, 79–82, 91–111; wing design, 91–111. *See also* Aerodynamics

Parnet, Claire, 65, 163, 185

Partial connections, 42, 86, 142, 161, 197, 202–203, 208 n.5

Partiality, 5, 41, 88, 126

Partial perspective, 41–42, 88, 212 n.14

Passivity, 137, 140, 151, 219 n.14

Pasveer, Bernike, 218 n.36

Patchwork, 86, 188–203

Patriarchy, 200; and TSR2, 138–140, 196. *See also* Gender

Pelly, Sir Claude, 55

Performance, 143; of centers of power, 151–153; of conditions of possibility, 148; ontological, 65, 147–148; of reality and fantasy, 147–148; and writing, 6

Performatives, 176–179

Performativity, 6, 11, 159, 174–182; narrative, 38–39, 144; narrative distribution and, 179–180. *See also* Performance

Periphery: performed, 153

Perrow, Charles, 217 n.27

Personal, the, 9, 38–64; civilizing process and, 43–48; social and, 47–48, 61–62; subject/object division and, 43

Perspectivalism, 21–23, 29–33, 36–37, 89, 121–125, 158–159, 177, 206 n.8; decision making and,

158–159; and subjectivity, 23; vanishing point and, 30, 121–124, 207 n.13, 215 n.13, 216 nn.17–18
Phantom, 145–146, 157
Pickering, Andrew, 5
Pictures. *See* Visual materials
Pilots, 31, 81, 83, 97, 104
Pinboard, 11, 188–203; complexity of, 198; as juxtaposition, 189; as methodology, 196–198; politics and, 198–203
Pinch, Trevor, 190–191
Place. *See* Spatiality: places of power and
Plain history. *See* History, plain; Narrative forms: plain history
Play, 221 n.24; and politics, 198–201
Pluralism, 214 n.1
Polaris submarines, 40, 172, 208 n.3
Policy: as narrative form, 56, 59–60, 68, 70, 220 n.18. *See also* British government
Political theory, 219 n.12
Politics, 39; grand narratives and, 200; interference and, 7; literatures on, 220 n.18; multiplicity and, 161–162; ontological, 198; pinboard and, 198–203; play and, 198–201; postmodernism, 8–87, 205 n.5; project-ness and, 7–8; quietism and, 7, 11, 161–162. *See also* Collusion; Criticism: political; "Politics"
"Politics," 143–162, 218 n.5, 220 n.18; as deleting "politics," 148–151; as opposed to "politics," 143–162. *See also* Decision making
Pollock, Griselda, 141, 216 n.18, 217 n.26 and 32
Postmodernism, 1, 7, 89–91, 164, 190–191, 198, 214 n.14; and expert knowledge, 1; modernism and, 2; pluralism and, 2; politics and, 87–88; skepticism, 1, 2. *See also* Reflexive modernity
Postmodernity, 53
Poststructuralism, 213 n.1, 218 n.1

Power, 220 nn.16,–17
Presence, 92, 94–96, 110, 113–114, 123, 160, 195. *See also* Absence; Oscillation: between absence and presence
Preston, 66, 71
Private-public distinctions, 38, 42–44, 58, 154–155, 220 n.19
Procurement policy. *See* British government
Production: versus consumption, 134
Project-ness, 87, 89, 175, 183–187, 197; arborescence and, 183–185; continuity and, 9, 71, 79, 87; interference and, 202; modern, 89, 114, 143; modernism and, 113; narrative and, 69, 71; and politics, 7–8; trajectory, 54
Public versus private. *See* Private-public distinctions
Punctualization. *See* Actor-network theory: punctualization in
Punctum, 218 n.33
Purity, 115, 118, 198, 215 n.7; versus impurity, 142
Puzzle solving, 77, 212 n.16, 216 n.22

Quasi-objects, 162
Quietism. *See* Politics: quietism and

Rabinow, Paul, 209 n.14
Radar, 26, 217 n.25
RAF. *See* Royal Air Force
RAF Cosford, 44–46, 51, 210 n.31
Rape, 132
Rationalization, 119
Realism, 17; produced, 127, 136, 217 n.32
Reality: fantasy and, 10, 147–148, 163, 217 n.32; performed, 6, 63, 65, 186; performed in text, 6, 7. *See also* Obduracy; Out-thereness
Reconnaissance, 169; TSR2 and, 13
Redemption, 189, 193, 211 n.7; into singularity, 195

Reflexive modernity, 1, 111–112, 190, 207 nn.10 and 12
Reflexivity, 7, 43, 45, 208 n.10; self-reflexivity, 45, 213 n.28; self-reflexivity as self-indulgent, 43
Reich, Steve, 126
Rembrandt, 134, 216 n.19
Representation, 214 n.11–12; text and, 6
Responsibility, 41–42, 90, 198
Rhizome, 1, 5, 6, 164, 175, 206 n.5
Rigor, 191
Rotman, Brian, 123
Royal Air Force, 66–69, 73, 82, 100–101, 106, 164–166; Operational Requirements Branch, 164–166, 211 n.9
Royal Australian Air Force, 25
Royal Navy, 73, 82, 104, 170–172
Russia, 101–102, 164–166, 214 n.7

SACEUR, 101
Said, Edward, 102
Sandys, Duncan, 83, 168, 170, 172
Scale, 9, 10, 164, 184–185, 214 n.5, 222 n.4. *See also* "Detail"; Narrative: scale and
Scenario. *See* Actor-network theory: scenario in
Science fiction, 219 n.9
SCOT. *See* Social construction of technology
Self-evidence. *See* Interpellation
Self-reflexivity. *See* Reflexivity: self-reflexivity
Semiotics, 8, 37, 48–49, 66, 92, 144, 206 n.1, 212 n.22, 216 n.20
Sexuality, 209 n.15; TSR2 as dangerous female, 138–140
Similarity: accountable, in decision making, 150; in culture, 70, 76; performed, 156–160. *See also* Difference
Simplicity. *See* Heterogeneity: heterogeneity/simplicity
Simpson, John, 33–34, 180–181, 220 n.18

Singleton, Vicky, 162, 207 n.16, 208 n.19
Singularity, 3, 4, 11, 53, 86, 88, 89, 115, 126, 142, 143, 183, 205 n.3, 214 nn.1–2, 223 n.5; assumed and enacted, 32–33, 65, 76; enacted in multiplicity, 194; incomplete, 111–112; redemption into, 198. *See also* Objects: and singularity; Oscillation: between singularity and multiplicity; Singularity and multiplicity; Subjectivity
Singularity and multiplicity, 3, 53, 86, 111, 115, 126, 142, 159–160, 163, 175, 183, 187, 193, 202; cyborg and, 3; of decisions, 159–160; fractionality and, 3, 4, 32–35. *See also* Fractionality; Multiplicity; Oscillation: between singularity and multiplicity; Singularity; Spatiality
Situated knowledges, 41–42, 46, 188
Size, aircraft. *See* Design, aircraft: size
Skill, 125–127
Social construction of technology, 84
Social interests. *See* Interests
Social shaping of technology, 70, 173; cultural bias in, 76. *See also* Technology studies
Spatiality, 53, 209 n.16, 219 n.8; cartographic, 23–25; Euclidean, 21–22, 173, 183, 185; narrative and, 174, 184; places of power and, 151–153. *See also* Scale
Speed, 92, 94, 99–102; depiction of, 29–32
Spitfire, 71–73
Star, Susan Leigh, 210 n.23
Stengers, Isabelle, 143
Story. *See* Narrative
Strathern, Marilyn, 95, 162, 188, 211 n.8
STS. *See* Technology studies
Studium, 218 n.33
Stuttering, 119, 185

Subjectivity, 9; aesthetic, 58, 61; Althusser's analysis, 50–51; cartography and, 23–25; centering and, 112–113, 115, 153, 159, 193; distribution of, 61 62; esoteric narrative and, 57–60; enacted as centered, 21; erotic, 31, 59, 64; ethical narrative and, 56–57, 59–60; female, 138–139; Foucault's analysis, 49; as fractionally coherent, 3, 4, 9, 197; history narrative and, 56, 59–60; interference and, 5, 59–61; interpellation and, 50–51; knowledge and, 4, 8, 186, 190–192; male, 10, 29–32, 40–41, 52, 122–123, 132–133, 138, 196, 210 n.31; as multiple, 2, 8, 59–61, 91, 143; and object division, 43, 45, 133, 175; oscillation and, 91; perspectivalism and, 22; policy narrative and, 56, 59–60; as singular, 25, 91; unmarked, 48. *See also* Agency

Supermarine. *See* Vickers Armstrong: Supermarine

Supreme Allied Commander Europe. *See* SACEUR

Symbolic interactionism, 222 n.6

Symmetry. *See* Method: symmetry in

Syntax, 18, 206 n.4. *See also* Coordination, strategies of

Synchronicity, 78, 84

System, 7, 185, 207 n.11, 213 n.28, 217 nn.25 and 27; narrative and, 9, 193; TSR2 as, 26–28, 85–86. *See also* Large technical systems; Weapons system

Tables: hierarchy and, 18–20, 184

Tactical strike and reconnaissance, 169–170

Taussig, Michael, 89, 188

Technical. *See* Technics

Technics: aesthetics and, 116–120, 131, 134, 140, 142, 156, 214 n.4, 215 nn.6 and 12; and agency, 125–127; deferral and, 131, 147; as

detail, 150, 215 n.6; nature and, 131; as serious, 131

Technological frame, 77, 212 n.16

Technology: as interpellating, 52, 64

Technology studies, 70, 213 n.1; grand narrative in, 86; narrative in, 70; project-ness and, 87. *See also* Actor-network theory; Social construction of technology; Social shaping of technology

Text: decline of, 189; visual material and, 10, 116–118, 121, 150, 196, 215 n.15, 218 n.38. *See also* Narrative

Thébaut, Jean-Loup, 199

"Third World," 164, 169–170

Threat, the, 102, 123

Time, 9, 11, 69–72, 77, 80, 85, 175, 183–185; arborescence and, 174; decision making and, 152, 161; deferral and, 110

Trajectory: project, 54, 55

Transcendence, 117–118, 196, 215 n.16; versus mundanity, 123–124

Translation, 99; as betrayal, 99. *See also* Actor-network theory

Traweek, Sharon, 38–39, 65, 70, 126, 208 n.7, 209 n.9

Treasury, the, 171–172

Trees. *See* Arborescence

TSR2: agency of, 122, 129, 133, 137, 216 n.17; brochure, 8, 10, 12–35, 116–142; cancellation of, 10, 40, 54, 143–162, 181; communications, 14; cost overruns, 149; ferry range, 25, 28; fuel system, 14; gender and, 10, 31, 137–139, 217 nn.29–30, 218 n.36; ground support, 27–28; history of, 163–187; as multiple or single, 12–35, 221 n.23; as mundane, 137; narrative of, 164–185; navigation, 26, 127; noncoherent, 33–35; as passive, 137, 151, 219 n.14; project, 54; project-ness, 87; reconnaissance, 13; refueling, 21, 135–137; sexuality and, 138–140; speed, 29–32; vulnerability, 129–131,

TSR2 (*continued*)
137; weapons system, 13, 26–28; windscreen, 128
Turnbull, David, 162, 209 n.11
Turner, Bryan, 199–200
Type 571: design of, 73–75, 79–81

Undecidability, 147
Unmarked subject, 48
USSR. *See* Russia
Utopianism: modernism and, 113–114

Valiant, 72. *See also* V bombers
Vanishing point. *See* Perspectivalism: vanishing point and
Vanitas, 42–44, 58
V bombers, 165, 167
Véhicule éléctrique, 218 n.4
Velvet revolution, 177
Vermeer, Johannes, 122
Verran, Helen, 162, 219 n.7
Vickers Armstrong, 71–85, 104–106, 113, 171–172, 211 n.5; culture, 77–79; merger with English Electric, 80, 172; Supermarine, 71
Viking, 72
Virtual object, 186, 217 n.23
Viscount, 68, 71–73, 211 n.5
Vision, 41–42; and touch, 216 n.19
Visual materials, 192, 196, 216 n.18, 217 n.26; text and, 10, 116–118, 196, 215 n.15, 218 n.38; secondary to text, 121, 150
V2 missiles. *See* Missiles: V2
Vulnerability, 137; versus invulnerability, 123–125, 129. *See also* Design, aircraft: vulnerability

Warton, 71, 110
Watkinson, Harold, 155
Weapons system, 75, 77–79, 110, 112, 168, 211 n.12; modernism and, 78–79, 212 n.17; P1.B as, 68; TSR2 as, 13, 26–28; Type 571 as, 73–74, 77; Vickers and, 77–79
Weber, Max, 47, 56–57
Weybridge, 71
Whitehall. *See* British government
Wigg, George, 144, 153, 219 n.14
Williams, Geoffrey, 33–34, 180–181, 220 n.18
Wilson, Harold, 145, 160
Wind tunnel, 108–109
Wing design. *See* Design, aircraft: wing
Wood, Derek, 55
World War II, 167
Writing: academic, 43, 190–193. *See also* Narrative

Zero point, 123
Zuckerman, Sir Solly, 154–155

John Law is Professor of Sociology and Science Studies at Lancaster University. He is the author of *Organizing Modernity* (1994) and (with Peter Lodge) *Science for Social Scientists* (1984). He has edited the following books: *Sociology of Monsters: Essays on Power, Technology and Domination* (1991); *Power, Action, and Belief: A New Sociology of Knowledge?* (1986); (with John Hassard) *Actor Network Theory and After* (1999); (with Wiebe E. Bijker) *Shaping Technology/Building Society: Studies in Sociotechnical Change* (1992); (with Gordon Fyfe) *Picturing Power: Visual Depiction and Social Relations* (1988); (with Brita Brenna and Ingunn Moser) *Machines, Agency, and Desire* (1998); (with Michel Callon and Arie Rip) *Mapping the Dynamics of Science and Technology* (1986).

Library of Congress Cataloging-in-Publication Data
Law, John
Aircraft stories : decentering the object in technoscience / John Law.
p. cm. — (Science and cultural theory)
Includes bibliographical references and index.
ISBN 0-8223-2812-7 (cloth : alk. paper) —
ISBN 0-8223-2824-0 (pbk. : alk. paper)
1. Entity (Philosophy)   2. BAC TSR 2 (Turbojet fighter planes)—
Miscellanea.   I. Title.   II. Series.
BD336.L39   2002   303.48'3—dc21   2001054711